OUTLINES

OF

A SYSTEM

OF

MECHANICAL PHILOSOPHY:

BEING A

RESEARCH INTO THE LAWS OF FORCE.

BY

SAMUEL ELLIOTT COUES.

"That which we call gravitation and fancy ultimate, is but one fork of a mightier stream for which, as yet, we have no name."

BOSTON:
CHARLES C. LITTLE AND JAMES BROWN.
1851.

CAMBRIDGE:
PRINTED BY BOLLES AND HOUGHTON.

TO THE

REV. ANDREW P. PEABODY.

MY DEAR SIR:

When I first mentioned that I had undertaken a Research into the Laws of Force, you encouraged me to follow up the investigation; and from that time you have constantly extended to me your sympathy, and have manifested a kind interest in the result of my labors. Thus you are intimately associated in my mind with the work, though it is but just to say that you are in no sense responsible for the correctness of any of my deductions. I dedicate the book to you as an acknowledgment of your kindness, and as a manifestation of my respect for the power of mind which has not only given you your eminent position in your profession, but is conspicuous in every department of learning to which your attention has been directed.

With sincere regard,

Your friend,

SAMUEL E. COUES.

ERRATA.

Page 107, line 30th — read, *descends to a lower level.*
" 163, line 8th — for *bury*, read *buoy.*
" 258, line 4th — for *their*, read *these.*
" 263, line 2d — for *common*, read *coming.*
" 272, note — omit the quotation marks in the second paragraph.
" 280, line 5th — insert *the* before work — next line, omit *the* before law.

CONTENTS.

CHAPTER VI.

CHAPTER VII.

CHAPTER VIII.

CHAPTER IX.

CHAPTER X.

CHAPTER XI.

CHAPTER XII.

INTRODUCTION.

THESE outlines of a System of Mechanical Philosophy are the result of a research made by the author, for his own instruction. In the research he attempted to separate supposition from fact, — the questionable from those things necessarily taken for truth by the laws of the mind. The inquiry took the form of synthetical reasoning, proceeding step by step, deducing the unknown from what appeared self-evident, taking nothing as truth on authority, however great the name which had given currency to the theory; thus shutting out all sources of error except such as resulted from the weakness of the individual mind. Arriving by this process at conclusions, the very opposite of those taught in the schools of science, these conclusions were tested by an appeal to the phenomena of nature. Both, the reasoning without the phenomena, and the phenomena without the reasoning, pointed to the same general truths.

Thus are presented opinions which appear to be supported both by reason and by observation, and presented in the order of their development. The writer feels confident, that, however defective may be his logic, however imperfect may be his statement of the facts of science, however crude may be his application of the truth to the explanation of phenomena, he has unfolded some principles, which, when distinctly evolved from mingled error, will change somewhat the basis on which mechanical philosophy now rests.

But, whether or not the opinions here presented are demonstrated to the minds of scientific men, it is admitted that many of the present explanations of philosophy are received as the supposed best possible explanations, rather than as settled truth, on which the mind confidently reposes. The conservatives in science may be successful in pointing out the errors of new theories, but they have other more difficult work, that of the defence of the old theories.

A reëxamination of principles supposed to be forever settled by the authority of a Newton may strike many minds as presumptuous. It will be seen, however, that the splendid conceptions of Kepler, made distinct and thoroughly demonstrated by Newton, are not assailed. The glorious truth of the harmony of the movements of the spheres, by which the astronomer can track out the orbit of the most eccentric wanderer of the heavens, stands forever, the reflex of the light of creation from the minds of the truly great.

It is the school which now determines opinions on philosophical subjects. Men are taught to believe rather than to reflect,—to receive opinions rather than to evolve them by the process of intense thought. A scientific education is a capital, the value of which is impaired by dissent. It would seem as if nothing further could be attained, as if philosophy had soared its highest flight; for theories thus pass from mind to mind, from generation to generation, as settled truths, and theories too, somewhat questioned or indistinctly comprehended, even by those who teach them.

To induce reflection, therefore, upon admitted theories,

cannot fail to do some service to the cause of truth; for, as it has been observed, "nothing tends more to the corruption of science than to suffer it to stagnate. The waters must be troubled before they can exert their healing virtues. A man who works beyond the surface of things, though he may be wrong himself, yet clears the way for others, and may chance to make his errors subservient to the cause of truth."

This work is not intended as an elaborate philosophical treatise, but as a rapid outline; and only those considerations are presented which are necessary to develop the opinions advanced. Much collected material has been laid aside for future use, if it be thought that the opinions set forth are worthy of additional attention.

PORTSMOUTH, N. H., Sept. 7, 1850.

CHAPTER I.

"FOR SINCE THE HUMAN MIND PARTICIPATES AS IT MAY IN THE FRUITFULNESS OF THE CREATIVE NATURE, IT DOTH FOR ITSELF BRING FORTH REASONABLE THOUGHTS."

IN all investigations of matter and force, we must divest them of all properties that are boundless, that are without fixed limits, determinate degrees. Thus, matter is not infinite in its extent nor in its capacity for motion; so of force, its action is not infinite; it has parts, or divisions, each with a fixed rate of power, and as a whole, its aggregate power is fixed,—meted out by Him by whose infinite power it was constituted. We need recurrence to this as the point of departure in our course, as that to which we can recur as something fixed, when the ocean on which we speed our way has no guide for our path over its vastness. There is one God, the ultimate Cause of all things, of whom alone can infinity of attributes be affirmed. This truth is here presented, not for its religious bearing, but from the necessity of recurring to it in its relations to the first principles of philosophy. The admission that His attributes alone are infinite in degree, is essential to distinctness of opinion with regard to matter and force. The human mind cannot comprehend infinity; any proposition, therefore, which involves the boundless, is unintelligible; decide it affirmatively or negatively, the mind feels no confidence in the result.

1. Matter has inertness as an essential property; it has substance and extension; it exists in atoms or ultimate particles, indestructible and impenetrable, between which atoms there is space; these atoms aggregated, constitute masses, between which there is space; these masses constitute worlds, between which there is space. Space is room or freedom for motion, or that in which matter exists and is moved. Motion is the act of the change of place of matter; it is the passing of matter from one part of space to another. No atom of matter can occupy the same place with another atom. Force is that which produces the motion of matter, or that which causes matter to act. Force cannot act where it is not.

These definitions are the usual, commonly received definitions of matter and force, and the resultant motion. We express in them all that can be known. An inquiry into their nature or construction embraces the element of the infinity of God who created them, and "plunges us at once into that deep which never yet was fathomed by human intellect." Newton thus speaks of matter. "We believe that God in the beginning formed matter in solid, hard, massy, impenetrable, movable particles, of such sizes, figures, and other properties, and in such proportion to space, as most conduced to the end for which He formed them; and that these primitive particles being solids, are incomparably harder than any porous bodies compounded of them, even so very hard as never to wear or to break to pieces; no ordinary power being able to divide what God himself made one at the first creation."

2. That which is moved and that which moves it, are not identical,—not the same thing; the one, matter, having in-

ertness as an essential property; the other, force, having as an essential property power to give motion. Wherever matter is, there is inertness; wherever force is, there is motion.

The belief of the separate and independent existence of cause and effect results from the constitution of the mind. It is what God compels us to believe. It is as inspiration. We think of God as existing separate from his creation, of our minds as separate from thought and perception; and so of force, as an independent power, as in no sense matter, but as that which gives motion to matter, as that which causes matter to change its place. This separate existence is recognized in the three great laws of motion, which refer to force as the cause of motion.

These words of Herschell very distinctly present the idea: "Force may be communicated to inanimate matter, it may be concentrated in the same mass by continuation of animal or muscular force. We thus learn to regard motion in matter as the effect and indication of force, and force may be defined as that which is capable of producing motion in matter, or of stopping and altering its direction when produced." Daniell (*Introduction to Chemical Philosophy*) says, "A body at rest would remain quiescent unless it were to receive an impulse from without," or objectively, and thus philosophy recognizes the inertness of matter. Growing from this idea is the doctrine of vis inertiæ. It is a term expressive only of the fact, that matter rests when unmoved, and moves when not at rest, and that time often elapses after the impulse, before motion is apparent. Strictly it asserts the impossibility of self-moving matter.

3. Matter which is in motion must continue in a state of

uniform motion forever, unless disturbed by the action of an external cause.

The same cause produces the same effects. This proposition, so far as it goes, is identical with the first of the three great laws of motion, which is, "A body must continue forever in a state of rest or of uniform motion, in a straight line, if not disturbed by the action of an external cause." If this be true of bodies both at rest and in motion, the proposition is true as regards bodies in motion, and as true when considered apart from the direction of the motion as it is when the law defines the line of motion.

4. All motion is in proportion to the force which produces it.

The degree of the effect is in proportion to the cause; in other words, add to or take from the force in action upon given matter, the quantity of motion is changed proportionally. The second of the three laws is: "Every change of motion produced by the external (objective) cause is proportional to the force impressed, and in the direction of a straight line in which the force acts." The proposition affirms the first clause of the law, and this is necessarily true independently of the truth of the latter clause.

We draw from this proposition the following as corrollaries: 1st, that force is divisible, different quantities or degrees of it acting at different times on the same matter; 2d, force is transferable, for otherwise, that is, without transfer of force, there could be no increase or diminution of motion.

5. The absolute condition of the transfer of force is its presence with matter not susceptible of motion by it.

The essential property of force being action, wherever it

is, it acts, or from its nature it is transferred from matter which cannot be moved to matter having space for motion.

6. The degree or velocity of the motion of a mass depends on the degree of force which acts upon each atom constituting the moving mass. It appears self-evident, that every atom, perhaps in a degree according to its element or kind, requires always the same force to move it with the same velocity, and the motion of the mass depends on the degree of force relative to the number of its atoms.

7. Every atom of the solar system is in continued motion, orbital and rotary.

Hence it follows, that it is force which determines the position of every atom, mass, and world, and determines their position positively in space. It also determines the relative position of all things.

8. Matter occupies but one place in space at the same time, and consequently no atom can have motion but in one line of direction at the same time.

Hence it follows, that all motion is orbital, rotary, curvilinear; for, all atoms being in curvilinear motion, no addition of rectilinear motion can change the curve into the straight line,—there will ever remain the element of the curve. The contrary opinion indicates infinity, and is unintelligible. The foundation principle of geometry is the measure of all angles by arcs of the circle, and the measurement of the circle or curve conversely by the inscribed straight lines or angles.

Hence the fixed relation between the diameter of the circle and its circumference, and the measure of motion by the diameter of the circular orbit. We assume that the harmonic motion of the heavenly bodies bears a certain proportion to the orbit, that this velocity has a fixed ratio to the diameter, and that this ratio mathematically determined is proportional to the square of the distance of the circumference from the centre.

The velocity of bodies moving in free space being in a fixed ratio to the orbit, it is the present force, the active principle, which determines the orbit. A change of the degree of force would enlarge or diminish the orbit. In all free motion, therefore, the motion is uniform and harmonic. Two atoms moving in the same orbit will move with the same velocity; in different orbits will preserve that relative position which is determined by the difference of orbit, and the difference of the line of direction; so that mathematically the relative position of any body can be deduced, given as the elements of calculation its orbit and direction of motion.

9. Rectilinear motion, or motion in a straight line, cannot be affirmed of any atom belonging to or in harmony with the solar system, or with the system of worlds. The horizontal line on the surface of the globe is curvilinear. So from any one point to another, in, on, and about the earth, the moving body takes the curved line. In the old philosophy, circular motion was deemed the natural motion; but the new philosophy conceives of circular motion as constrained by the operation of conflicting forces.

10. The supposed perpendicular descent of a falling body is a curvilinear motion. This is not practically admitted, but mathematically deduced; the fall is termed a "cubical

parabola." The supposed attraction of the earth does not draw in straight lines, falling bodies preserving ever their rotary motion.

11. Bodies in rotation with the earth, on receiving additional force, must either take a higher level of rotation, or use the force received in additional motion in a new direction.

12. In the act of falling whereby a lower level of rotation is assumed, force must be transferred from the falling body when its motion is suspended.

The motion of falling is added to the rotary motion, thereby giving action to the present force. This motion being suspended, only rotary motion in a smaller orbit remaining, force will be transferred. Hence arises spare force of descent. The spare force of descent will be measured by the degree of descent. Hence the spare force of falling bodies is measured by the square of the time of descent, equal times giving equal distances of descent. Thus most distinctly is presented the law of falling bodies, or motion begetting motion, which, indistinctly understood, caused the great controversy between the schools of Des Cartes and Leibnitz, and which to this day has obscured mechanical science, by making the occasional abnormal motion of falling bodies the element of calculation of harmonic motion.

13. The motion of the heavenly bodies being harmonic with fixed velocity according to orbit, their velocity is the immediate measure of the velocities of retarded or accelerated motion; that is, the abnormal motion is detected and

measured by the normal, the motion of bodies in confined space being determined by the velocity of bodies in free space.

Time is the indirect measure of velocity; the motion of the earth which measures time, the direct measure of velocity. Hence all velocity is known by and measured by the rotation of the earth. A body which moves thirty miles in one hour moves over the given space isochronously, or while the earth has moved one twenty-fourth part of her rotary circle. Time, therefore, has no reference to force and motion so as to increase or diminish them, and motion refers to time only as indicating the extent of the motion of the earth.

14. There is beside the consentaneous motion of the atoms of the mass, or the progressive motion of the mass, an atomic or molecular motion of every atom composing the mass.

This proposition will receive its most perfect demonstration by the facts and phenomena of nature; but is believed from analogy, that the same law governs the minute as well as the extended, — also from the fact that there is space between all atoms, that force is primarily in action on the atoms. If force be present with atoms, and there be space for motion, motion is affirmed. We leave this proposition mainly upon these general statements, recurring, in this place, to one most significant fact: "The expansion of volume of small bodies by dilatation from heat, may be obtained without sensible error by trebling the number which expresses its increase in length." Daniell's *Introduction to Chemical Philosophy*. The trebling of the diameter, it will be perceived, gives nearly the circumference of the circle of motion.

15. Force transfers itself from atom to atom without appreciable time. Time being measured by the action of force upon a given quantity of matter, the transfer of force without matter is therefore without time; or with a reduction of the matter, velocity is increased; annihilate the matter, and time is no longer to be affirmed of its passage. Thus attraction of gravitation is spoken of "as a force which is transmitted instantaneously," and so of light, heat, galvanism, &c. So far as they pass in appreciable time it is because they (the forces which move) are connected with the material; the essential quality of pure force unembodied is change of place without appreciable time.

16. Force is diffusible only by contact of the moving atom with another atom. This appears from the fact that force is transferable only when it cannot induce motion, and while space remains there is room for motion. Therefore, the motion of a body must be after a lapse of appreciable time from the contact of the atoms. Consentaneous motion of the mass is from the equal diffusion of motion through its particles. These particles have room for motion before impingement; therefore, the mass moves in appreciable time from the contact of the atoms.

This is finely illustrated by a row of suspended ivory balls through which force is communicated in appreciable time by the measured motion of the balls, according to their distance and the degree of force. It is the foundation of the doctrine of vis inertiæ. It is not impossible that the distance of atoms constituting a mass may be relatively determined with regard to the distance of atoms in another mass. It will be found that, the more porous the body, the longer is the lapse of time between the impingement and the motion of the atoms as the mass; we instance equal weights of atmospheric air and of lead. The force is communicated instantly, motion results instantaneously, but time is predicable of the passage of atom to atom, to diffuse the motion equally that it may be consentaneous.

17. Force acts independently of direction; on its transfer, its former line of motion does not necessarily imply a continuance of the same line of motion. A body in motion must continue forever in a state of uniform motion, unless disturbed by the action of an objective cause. The force will be transferred upon impact, and the body receiving it will move in the direction in which it is free to move; or, if the body which obstructs motion be not movable, the force, not being transferred, acts in the opposite direction.

The resultant motion of impinging bodies is demonstrated by the mathematical figure, the "polygon of force," and is equally demonstrable by the diagram showing the contact of atoms having surface; so, too, the angle of incidence and reflection, when a body impinges upon an immovable mass, shows that the resultant motion is determined only by the free space for motion. If two flat surfaces meet, they can be separated only in the opposite direction, if the separating force acts equally on the surface. And so through more complex movements.

We refer to facts of common observation; the rotary motion of the windmill by the horizontal impingement of the wind, — the billiard ball, — the rise of the wave. On this action of force, independent of direction, depends the application of force in mechanical combinations, by which the force of falling water is conveyed anywhere, and acts in any direction. Force does not act where it is not; but the mechanic, availing himself of the laws which govern its transfer, leads it to the loom and opens for it the desired line of motion only.

This doctrine of the transfer of force is one of the foundations of the supposed action and reaction of force; every instance of reaction can be directly traced to the transfer of force.

18. Transfer of force, which does not give the body to which it is transferred consentaneous motion of its atoms, gives vibratory motion to the parts of the mass, or increase of molecular motion, the motion induced being of the same degree as the motion that was suspended. The force being transferred, and motion being in proportion to the force, the vibratory or molecular motion must be the same; or, if the motion of the recipient of the force be retarded, an equal quantity of its force is also turned from the consentaneous motion, so that the resultant motion, internal to the mass, will be doubled; that is, the internal motion is the aggregate of the suspended motion of both impinging masses. Thus, the motion of the tongue of the bell, and of the bell, jointly increase the loudness of its tone; and other instances of this result present themselves to every mind. Hence the concussion and destruction of two balls in opposite motion, which would not result from the motion of either upon the other at rest.

Vibratory motion being the unequal or rather unconsentaneous motion of the parts of a mass, if there be not a gradual increment of motion from the surface to the centre, there will be nodes or points of rest. Thus musical chords manifest certain points of rest.

Vibratory motion being determined by the degree of force,

if the same quantity of matter be moved the same distance, the times of vibration will be the same; if the distance be diminished, there will be an increase of velocity. Tone being dependent on time of vibration, an increase or diminution of force changes the range, preserving the tone. Isochronous vibration is, therefore, a branch of the law which induces the same quantity of motion by the same quantity of force. It follows, too, that there can be no vibratory or reciprocal motion in free space.

19. Force is indestructible. This appears from the general consideration, that what God formed at the beginning no power can destroy. It is imperishable, too, as is evident from its transfer from atom to atom and from mass to mass, and from the state of equable motion, which bodies actuated by a given degree retain. It is also one and identical. That which gives motion is force; if it have different essential properties, these properties cannot be detected, as force is known only by the motion it gives to matter.

There are not differing qualities or kinds of force, which give, — this kind, circular motion to bodies, — this kind, a straight motion, — this kind, motion in a contrary direction; nor is there a planetary force, and a molecular force, and an oscillatory force, and a directly acting force; but the one force may act on any element of matter, moving it in any direction, which idea will receive further illustration in the following chapter.

In the research made, the general propositions were extended much further; but up to this point only were they carried with that distinctness which should characterize such propositions. Others remain therefore for further examination.

It will be observed, that an inquiry like the preceding has reference to force mainly as in action, giving to matter place, position, and motion in reference to space,—actual, and not apparent or relative motion. We try to consider relative motion only as indicative of true motion, and this as indicative of the laws of force. Thus the ascent of water in the capillary tube is regarded, not as motion in relation to the tube so much as a new rotatory orbit of the elevated water. The log line which is veered from the ship has apparent motion of its own, but the actual motion is that of the ship. This distinction we would recall to the mind of the general reader.

CHAPTER II.

"WHEN WE HAVE THE DECREES OF NATURE, AUTHORITY GOES FOR NOTHING." — *Galileo.*

THE errors of mechanical philosophy, if there are errors, arise from the assumption of gravitation, or the attractive power of matter; from the application of the law of the motion of falling bodies, to the uniform motion of bodies remaining in one determined orbit; from the belief that rectilinear motion is the natural motion, and that the curvilinear is a constrained motion induced by conflicting forces; and from keeping out of sight the intense motion of every atom in its rotation and revolution with the earth, which, from its greater comparative velocity supplies the governing or controlling motion, and by reference to which alone, incidental, retarded, or accelerated motion, is to be understood or to be explained.

In the illustration of our views we shall follow no artificial arrangement, but present our thoughts in the order in which they occurred from the natural connection of one class of facts with another class of facts, as this will throw more light on the peculiar views than any predetermined order of investigation.

Our first attempt is to illustrate the idea of the identity of force, — to prove that the motive power of nature is one, acting under uniform laws; that force, whether it form the dew-drop or marshal the "hosts of heaven," whether it manifest itself in the flow of the tides or of the purple stream of life, whether in the flash of lightning, or in the sweep of the bird with motionless wing, is ever the same principle; that its divisions and subdivisions, its ever-varying names in science have obscured the law of its action, and thrown a mist over philosophy else apparent, from the ever-changing and conflicting theories.

Of the nature of force as well as of the matter it moves, as we have said, we are entirely ignorant, and from the constitution of the mind we shall forever remain ignorant. But the laws of its action given to it from the beginning, and ever enduring, may perhaps be distinctly traced out and understood. Thus, the force which moves the spheres acts under the intelligible law, that velocity is in proportion to the area of the orbit, that with the diminution of distance from the centre velocity increases, with the increase of distance velocity diminishes; so that all bodies revolving in free space describe equal areas in equal times. This is a property of orbitual revolution, — in other words, it is the property of force. It is a fixed law, subject to no change. Aberrations and perturbations are temporary and oscillatory; the mean time of revolution never changes.

The revolution of the heavenly bodies is accompanied by rotation on their axis. This connection of the primary with the secondary movement arises from another law of force, — from another property of the power which imparts motion. The two movements appear to have an opposite character. In the one, as we have seen, velocity increases by the diminution of the orbit; in the other, velocity becomes lessened as the rotating matter approaches the centre of rotation. How slow the rotation of the central parts of the earth, while at the equator the velocity of the surface is more than one thousand miles an hour!

From the nature of the rotating sphere there is one fixed ratio of increase of the velocity of its parts. Given the velocity at any distance from the centre of rotation, the velocity at any other distance can be determined. Its ratio of increase is proportional to the increase of the area of the circle described. We have the same measure of increase outward in rotation, that we have inward in revolution. The primary and secondary movements are under the same general law. Besides, there is also a fixed ratio between the velocities of both movements, for such is the far pervading law of nature.

If the orbit of any planet were enlarged without additional force of propulsion, it would move, in its new orbit, with decreased velocity. Its motion would not be harmonic; it would not describe the equal area of planetary motion. To bring it into harmonic motion it would require additional force, and this additional force needed would be measurable by the increased area of orbit. If, on the other hand, the orbit of any planet were decreased, with the same propelling force its velocity would be increased, and it would be out of harmony with its associated worlds. With equal velocity it must part with, or transfer, a portion of its force. This degree of spare force would also be measurable by the reduction of the area of the orbit. The degree of force required or imparted is not measured by the increase or diminution of the circumference. The orbit does not measure velocity. It is a path of motion, continuous, without beginning or end. By a fixed law, the required force is determined by the area described, which increases and diminishes in a higher ratio than the length of the circumference which encloses the area. Thus, by the enlargement of a planet's orbit, it would need added force to preserve the harmony of planetary motion, and by its diminution it would impart force, and the force added, or given up, would be measurable by the change of area. In the words of Herschell: "The law of the areas determines the actual velocity of the revolving body at every point, or the space really run over by it in any given portion of time." If the velocity be thus determined, so is the present force which determines the velocity.

This is equally true of the velocity and force of rotation. If a mass at the surface of the earth be elevated, it gains thereby a superior level of rotation; its orbit is enlarged,—it requires additional force. If the mass fall, thereby decreasing its orbit of rotation, it has the spare force of descent; it requires less for rotation, and the falling body imparts force. The force received for elevation, the force given out by depression, is measured in one case by the increase, in the other by the decrease, of the area of the circle of rotation. Though in one view it appears as if there were the converse action of force, yet the general law is apparent, of the increase and diminution of required force for any area of orbit, whether acting outwardly from the centre or inwardly to the centre of revolution.

The identity of force, acting in the primary movement of the spheres, and in the secondary movement, is apparent, thus considered, and a general, far-reaching law is most clearly indicated.

The force which moves the different parts of the earth in rotation differs much in intensity. How small, comparatively, in the central masses at the axis of rotation; how great, on the surface of the earth at the equator; how different the force moving the surface of the earth at the equator from the force moving the surface of the earth at the poles! Yet, the weight of falling bodies, or their spare force of descent is nearly the same at the equator that it is at the poles.

The spare force of descent is most unquestionably greater for an equal change of orbit at the equator than at the poles; but the change of the area of the orbit is far less by the descent of one foot at the equator than it is at a distance from the equator. As we go from the equator the force of rotation lessens, the area of the orbit lessens in that ratio by every foot of descent, so that the increased ratio of the decrease of area compensates for the decreased force of rotation. Thus have we an equal "spare force" of descent on the same parallel of latitude, slightly increasing as we go from the equator, as by the spheroidal form of the earth the decrease of the area is more rapid by equal decreases of diameter, in proportion as the earth changes from the perfect sphere to the oblate spheroid.

There is not, then, one law of force for the revolution of the spheres, and another for their rotation, and another for the changes of level of the masses composing the spheres; and we think that we can show that the law, by which the spheres and masses are moved, also directs and governs the motion of the atoms composing every mass and sphere.

The spheres lay spread out before us as distinct objects of conception. From the immense extent of space they occupy, and the almost boundless range of their motion, the law of force indicated by them becomes intelligible, and is the object of precise reasoning and of mathematical calculation. If we would understand the minute, we must throw upon it the light gained from the extended. "There is a certain character or style, if I may use the expression, in the operations of Divine Wisdom; something which everywhere announces amidst an infinite variety of detail, an inimitable unity and harmony of design, in the perception of which philosophical sagacity seems chiefly to exist;" or, as Newton expressed the same idea, "Nature will be very conformable to herself, and very simple." Thus, no less exact, no less fixed by determinate law is the motion of atoms, and, in the examination of the minute, this idea of the extent and universality of primary laws should be ever present to the mind.

How unlike the simplicity of nature, her directness of action, the far-reaching nature of her laws, is the array of opposing and conflicting forces, by the clashing of which we are taught that the balance or equilibrium of things is preserved! To return for a moment to the spheres, — there is the impelling force, the repulsive or centrifugal force, the attractive or centripetal force. Of what use are these? "I am persuaded," says Plato, "that if the earth is placed in the middle of the heavens, as they say it is, it stands in no need of air or any other support to prevent its fall; its own equilibrium will keep it up. For whatever is equally poised cannot incline to either side, and consequently stands firm and immovable, this I am convinced of." If the centripetal and centrifugal forces balance each other, they induce no action; give the earth the force for harmonic motion according to her orbit, she needs neither to keep her in the track; there would be no danger of her flying off in a tangent, or of her falling into the central fires of the sun; evenly balanced with them, she would be equally balanced without them.

Why these conflicting forces were created by philosophy

is very apparent. They follow from the strange assumption that nature abhors the orbitual or curvilinear motion; that, in the words of Herschell, "A straight line, dynamically speaking, is the only path which can be pursued by a body absolutely free, and under the action of no external" corrective "force." Centrifugal force is therefore not exactly a force, but the tendency of nature to escape from the confined curvilinear motion, and to get back the sphere into the straight line of motion. Philosophy looks upon the myriads of rolling spheres as in unnatural, constrained movement, and there must needs be conflicting forces, to form the grooves and channels which hedge in the planets, and keep them to their orbits, when they would so rejoice in free straight motion to the bounds of space.

Whewell says that, "In cosmical phenomena, every thing in proportion as it is referred to mechanical principles tends to simplicity, to permanent and uniform forces, to one common property. In magnetical and chemical appearances, the application of mechanical principles leads only to a new complexity which requires a new explanation, and this explanation involves changeable and variable forces, gradation and opposition of qualities." That is, plain, simple, mechanical laws apply only to the enlarged action of nature; in her recesses there is confusion, unintelligibility. For the want, therefore, of fixed general laws in the minor mechanism of nature, or for the lack of comprehension of those laws, the department of "mechanico-chemical" sciences is cut off from general philosophy, and managed by a great array of conflicting forces, such as heterogeneous attraction, homogeneous attraction, capillary attraction, various forms of repulsion, elasticity, cohesion, chemical affinity, current affinity, and the forces of the imponderables. Every judgment in this department is particular. "There is no enlargement of view to general propositions." We see deep thinking, "but the conclusions arrived at, even though clearly expressed, are intricate and obscure." There is not in the teachings of mechanico-chemical science the prominence, the boldness, the exactness, the simplicity, which characterize the works of nature, whether her force be exerted on the atom or on the world.

Imponderables are so called because they are not the subject of weight. They impart no force by the act of descent, but for this reason they are not proved to be immaterial, without matter. We read, (*London Encyclopædia*) that "a stream of electric fluid issuing from points, possesses force sufficient to counteract the power of gravitation in light bodies"; of course the imponderables could counteract the gravity of any matter which was their vehicle, or rather which they moved. Gravity being the unused force of rotation, given out on the arrestation of the motion of the falling body, they have none of this spare force; they cannot be arrested; they consume all their inherent force in their rapid flight. It is only by the stopping of the descending motion of falling bodies that they give out force or indicate weight. We know not of pure disembodied force, and believe that the imponderables are the combination of force and matter, the force in intense degree compared with the matter; that the element of matter, and its quantity in the possession of force, makes the distinctive character of the imponderables, and that force is thus fitted for especial functions in the economy of nature. This view does no violence to any system of philosophy, and appears demonstrable by the views we have taken in the preceding propositions.

First, of heat. Heat has been regarded as the repulsive force, as the diluent of force, keeping it in due bounds. The same philosophers who would set aside mechanical action with reference to the mechanico-chemical sciences, at the very outset bring back the discharged mechanical laws. They need a pendulum or balance spring, — something by which to regulate and control the mixture of forces. Heat is elected to this office of antagonism. The perfect law of force needs not this restraint. But there is a peculiarity of the action of force under the condition of heat, which has been generally referred not to its nature, but to its connected matter. It seems fitted thereby more especially to induce corpuscular motion, and the heat acts plainly and distinctly in giving consentaneous motion to masses; yet most usually its effects are traced in an internal action.

We notice in this place the difference between solar and radiant, or reflected heat. The passage of solar heat through a lens shows that force then has under its control matter, the atoms of which have extent of surface, as will be explained when we refer to light. Radiant or reflected heat is not reflected or bent by the passage of its rays through a glass lens, but is absorbed by the glass, which becomes heated, and continues so until by conduction the heat is equally diffused. It appears, then, as if by the impact it deposited a portion, at least, of its accompanying matter, and, the matter being deposited, the heat in passing through the lens is absorbed. The force has taken its connection with other matter. Radiant heat is then force partially divested of its matter; partially, we say, because in its passage through a lens of rock salt, it again obeys the law of the passage of light and heat through a transparent media. A secondary radiation of heat still further deprives it of its matter. Thus we come to the conclusion that these differences are caused by the deposit of matter at the impingement of its rays. It shakes off, as it were, its impeding embodiment, that with unimpaired energy it may give its strength to the corpuscular movements.

The action of heat is readily explained. Every species of matter has that quantity or degree of force necessary to maintain its atomic structure. This is called latent heat, and is latent force. Now with the normal degree of force in molecular action, another degree of heat or force can be received and made latent in the corpuscular action without apparent change of structure. This is specific heat; another added quantity becomes sensible or diffusible heat. It is beyond the normal heat which its structure can bear, and it swells or enlarges. Increase the heat still further,—the molecular action increases, the orbit of circulation enlarges, and the solid becomes fluid and the fluid aeriform. Hence heat acts repulsively by the very law which fixes the orbits of rolling spheres, that of increasing circles with increasing force. We have already referred to the fact, that the volume of the dilating body increases three times in small bodies for its length of increase, indicating the ever circular orbit of force. Latent heat is therefore normal molecular action, capacity for heat, that which a body can bear, and use internally without change of condition; sensible heat shows itself by dilatation and expansion, by the increase of the circular motion of its atoms, and by its transfer.

Daniell says that the polarity of heat is as certainly demonstrated as the polarity of light, though the experiments which demonstrate it are of a more delicate nature. Polarity is circular motion, or rather the indication of circular motion. It is the direction of the orbit, and, applied to matter through which force circulates, the poles are the anode and the cathode,—the direction of the current which passes; or as Whewell describes it: "Polarization indicates opposite properties in opposite directions, so exactly equal as to be capable of accurately neutralizing each other; that is, the same force passes and repasses circularly with an even power."

Heat is always accompanied with a greater or less degree of light. This is certain with reference to free current heat. Latent heat and conducted heat are not the subject of examination; but yet as the breaking up of cohesion manifests light with the evolved heat, it appears almost certain that light is the essential condition of force manifesting itself as heat. What is light? Not inert sluggish matter, not the disembodied force instantaneously diffusing itself, but the combination of the two. This is the reason of the endless disputes about the materiality and immateriality of light; both views are partially right, both are wrong. There are indications that will support either side of the question, but the phenomena collected together result in the union of the two. It is force which gives it its transit, its dynamical laws; it is matter which strikes the eye and forms the picture there. It is what force transports by which the world is painted in all its glory. It is *not* "a propagated quality of motion, extending in right lines in all directions from the point from which it emanates;" but something is borne along on the wings of force so subtile as to be felt only by the most delicate tissue of the animal system, the retina.

The materiality of light has the support of Newton and the great philosophers of his day, and of many subsequently.

Its materiality seems demonstrable from its being absorbed by the vegetable tissues, giving them color and strength of fibre which no "propagated quality of motion" can give. It is inherent in the diamond; it can be collected from the sun's rays by solar phosphori, and when emanated, the supply can be again received by a fresh exposure to the sun; it is evolved from decaying bodies and from the living animal organization; it obeys the laws of motion, of known moving matter, its line of motion is bent, and when not absorbed by the body on which it impinges, as the ponderous missile on the ice or water, the angle of reflection is equal to the angle of incidence.

Light also changes its line of motion in passing through a denser medium. This change of direction is only at its ingress and egress. It is a surface change, determined not by the medium, but by its transit within and without. May not this be the theory of the change? The atoms of light, being matter, have extension of surface; they strike the concave or convex lens on their edges as they enter, at an angle, not with their surfaces parallel to the glass; on the egress one edge escapes while the opposite is yet resisted by the denser medium; to use a popularly technical word, they are *canted* by the convex towards a centre, by the concave from a centre or focus, both on entering and leaving the glass. It appears, too, as if light were composed of atoms of different extensions, as by the prism these atoms are separated by their greater or less impinging surfaces. The fall of a flat piece of wood angularly on the water which would be turned, illustrates the idea; in the case of the atom of light, the concave or convex glass would make the angle. Does not this also explain the angle of reflection? It is remarkable, too, that the refractive powers of media are in proportion to their densities; water acts more powerfully than air, and the power of water is increased by its having salt in solution. Bring to the mind, too, that the change of direction is at the surface only, and we must impute it to the manner of its entrance by surface change. A wave of ether or a "propagated quality" of motion by act of impingement would accommodate itself to the angle of impingement, as do the waters of the ocean to the projecting cliffs of the shore. The theory illustrates the phenomena, even if it be drawn from the imagination only.

But it will be thought that our view is too mechanical and rough to be affirmed of light; for how exceedingly minute must be those particles which in myriads cross and recross without impingement! Space is indeed incomprehensible in its minute divisions, as in its most extended bounds. It seems to flow out on either side, — in wave-like undulations, — upon the confines of infinity, with which it blends without distinctness of separation. Obscure as it is to us there is about the atom room, fixed laws, accuracy of motion, and to higher beings the laws of God, operating in a mere point, may be as palpably distinct, as to us is the waving of the branches of the tree that bends before us in the summer's wind; and because the motion of the atom is obscure to us, must we throw upon it the reflected light from the phenomena which nature, in her enlarged sphere, places before the eye. We understand the minute only as we comprehend the extended; for they both have the one God over them, whose law is universal, unchangeable in reference to the atom and to the sphere.

We advert but to one more of the phenomena of light, — its polarity or the circular motion of its atoms. This is the stamp of force, its great characteristic, ever denoting its presence. We will describe polarization in the words of Daniell. "It may assist our comprehension of the phenomena to illustrate them by a rough analogy; a ray of common light as it is emitted from a self-luminous body, we may conceive to revolve upon an axis coincident with its own direction, as a cylindrical rod may be made to turn, or which comes to the same thing, the reflecting or refracting surface may be made actually to revolve around the ray as an axis preserving the same relation to it, and no change in the phenomena will be perceived." We add, that the rolling earth flying in her orbit would be an illustration perhaps more apposite.

We proceed to the examination of electricity. Franklin, with the sound discrimination which characterized his mind, expressed the truth in relation to it. We will quote his language, substituting the word "force" for "electrical fluid." "The opposite states of electrical excitement depend upon the increase or diminution of *force* relatively to the two bodies." These two states have been called the positive and negative. When two substances are rubbed together, sometimes the one and sometimes the other is excited. All substances have been divided into electrics and dialectrics, the one the conductors, the other the non-conductors of electricity; but subsequently it has been ascertained that all substances are conductors, their conducting powers differing only in degree. Hence we have *not* conductors and non-conductors, but substances differing in being *good* or bad conductors; the metals, water, animal and vegetable bodies on the one hand, and oil, glass, dry air on the other. There is in the extremes a striking difference of atomic structure, as for instance, in iron and spermaceti.

Any dissimilar substances rubbed together with proper

precautions are capable of exerting the power, one of them becoming excited. The effect is produced by the application of force, nor does it change its nature. When confined, it is force; when again set free, it is force. Electricity is thus described: "It is freely evolved by the mechanical action of heterogeneous substances; still more abundantly by chemical action; it is a product of animal organization; it is capable of evolving heat, and heat is capable of evolving electricity. It is capable of acting on bodies in opposition to gravity, and is capable of the most energetic action on its own constituent atoms. Every form of matter may be excited to its action, and it may be transferred from one portion of matter in which it has been called forth, to another previously in its natural state." We could not more distinctly describe force. Electricity is force existing under some peculiar conditions. Its polarity is one of its most prominent phenomena. It can be insulated, or a channel formed for its circulation; it can be kept a prisoner, or its path opened that it may return to the earth, and add the strength of its feeble stream to swell the torrent of force which bears along the mighty orb in her rolling path.

Differing from galvanism, which appears as a continuous stream of force, electricity is, as it were, a collected mass, a sphere revolving on the surface of the glass. It does not seem to affect the molecular action, but to pass through the aggregated atoms as free, revolving in its own orbit; for, if you break the cylinder which contains it, the electrical spark is found in one of the parts only. Electricity, too, gives another striking demonstration of its nature. We quote: "If the electricity collected be distributed between balls of different diameters, it will be found that, the smaller the sphere, the more intense the action." Such is the law of force, wherever it exists; there is circular motion, equality of diffusion, the velocity increasing with the decrease of the diameter of the orbit. In the excited rod, too, the action is more intense at the ends than at the centre; for the law is that of increased speed with the narrowing diameter of the elliptical orbit. Electricity gives to matter no attractive power, — as it partially escapes from its receptacle, it draws or repels other bodies towards or from its receptacle; and whether attracted or repelled, the moving body determines its motion by its own state of excitement. If equally excited, it adds to the circulating force; if less excited, it absorbs a part of the circulating force. No attraction of matter can be affirmed of this fitful, spasmodic action of bodies floating in the vortices of revolving force. How beautiful the provision by which the unexcited body advances to the excited, then in contact receives its burthen to be conveyed to other bodies, so that the resultant motion of vertical force is the equal distribution of that force, — an equilibrium preserved not by opposing forces, but by the very nature and laws which govern the action of force.

Galvanism is the result of the breaking up of existing structures, thereby freeing the force which bound their atoms together. Would any one ask what is the cement of the building when, by its destruction, the earth is covered and the air filled with lime? It is the force of cohesion made current, which constitutes galvanism. The condition of its being "current affinity" is decomposition, and a conducting circuit or orbit in which it is to move. "Now," says Daniell, "it is curious to observe how chemical affinity in all these experiments waits upon the conducting and collecting power; the strong attraction of zinc for oxsulphion is held in complete check till a passage is open for the circulation of the force ; the force which circulates must be equal in all parts of its circuit, it stops with the slightest break in the continuity of its conductor." Urged sometimes by the intensity of the charge, it leaps through a short space of air, and then with its flame arched upward, proclaims the tendency of force to assume the circle, whose circumference is proportional to the degree of the force which revolves.

Dr. Farraday has proved that the contact of dissimilar metals is not necessary to the generation of the current, and that the force is always in exact proportion to the amount of chemical affinity. With a knowledge of the laws of force, this fact needed not the demonstration of experiment; for the force of cohesion liberated must measure the current force ready to act in new combination, or to pass for examination along the conducting wire to the matter it is to bring together in newly built structures. How beautiful, too, the analogy between the chemist and the mechanic ; the former, commanding for his work the force once used for cohesion, and again directing its strength for a similar purpose; the latter, using as his property the force of a moving world, directing it in the circuit of his machinery, that it may destroy structure and build other structures that minister to human wants. This very analogy has been thus traced out. "The analogy of the transmission and direction of mechanical force, may perhaps assist in the formation of a clear idea of this influence of bodies," [current affinity.] "Every one is familiar with the modes by which the muscular force of animals, the elastic force of steam, &c., is led by the solid matter of levers, cords, and wheels to distant points from its source and there set to work, somewhat in the same way we may conceive that the force of affinity may be directed to distant points through appropriate conductors. But its journeyings must be in a circle, and the arrangement must be made in such a way that the impulse may return to the point from which it set out; it must *circulate.*"

"In no case," says Farraday, "is there a pure creation of force, a production of power without the exhaustion of something to supply it." Of course it has excited the wonder of chemists, that two pieces of metal, producing no change in their states, should be an inexhaustible source of power. The wonder ceases when the distinction is made between current affinity brought into action, and continuing in action in its orbit, and the current affinity from broken cohesion, returning after its circuit to rebuild from the elements which had been torn asunder, at the broken orbit between the positive and negative pole. The one is seen in the action of the magnet,—the other in the passage of the galvanic current, passing from the work of destruction through the wire to the work of re-creation, or at once passing through bodies in chemical affinity.

Before we proceed to the examination of the union of the magnetical and electrical phenomena and the action of the magnet, we will explain our views more fully of the effect of the galvanic current in changing the condition of bodies on which it acts. Polarity is the positive and negative pole, or the direction of the current force. The wire is the conducting circuit, between the poles of which is the substance acted upon. "Now the idea of polarity," says Whewell, "involves the conception of opposite properties in contrary directions; for example, attraction and repulsion, darkness and light, synthesis and analysis." That is, force revolves, the current in its passage one way repels, in the other attracts; in one way gives darkness, in another way light; it breaks down structure or rebuilds structure, for in one way force is taken from the body and from the other pole it returns. In the one case, for instance, it resolves water into its constituent gases, in the other combines the elements of water into that fluid. This proves that our idea of polarity is right, and opens the mystery of chemical action generally; for the same result is had without the wire, — the transfer of force. Farraday expresses the idea in almost the same language. "Chemical synthesis and analysis must be conceived as taking place by virtue of equal and opposite forces, by which the particles are separated or united," that is, by the transmission of force.

We continue to quote: "These forces, by the very consideration of their being polar, may be transferred from point to point, and thus we have a positive force active at one extremity of a line of particles corresponding to a negative force at the other extremity; all the intermediate particles neutralize each other's action." This idea was introduced by Prout, confirmed by Davy, and fully illustrated by Farraday; and for its full comprehension it needs only that we discharge the unmeaning expression, 'negative force,' and consider the phenomena as indicating the transfer of force in its circulating orbit.

In our view of magnetism, we have the labor lessened by the acknowledged fact, that current affinity, or galvanism, or electricity, is identical with magnetism. It is so decided by competent authority. This was first announced by Œrsted in 1820, and proved by a series of experiments. Franklin long before this had induced magnetical polarity in fine needles by passing through them a current of electricity. The disturbance of the polarity of the ship's compass is also a well known fact.

We call attention to this phenomenon: place the needle over the conducting wire; it lies across the path of the rushing force; beneath, it lies again across the marked pole in the contrary direction. Dr. Farraday succeeded in producing a revolution of the needle about the wire, and of the wire around the needle. The principal effects of terrestrial magnetism "may be imitated by distributing a wire around the surface of an artificial globe, in a spiral direction from the equator to the poles, the two extremities being turned inwards, brought out at the two axes, by which the connection may be made with the battery. A magnetic needle properly suspended in different situations near the globe, will arrange itself in positions perfectly analogous to those assumed by the dipping needle in the corresponding regions of the earth." The action of conducting wires rolled in the form of the flat spiral, produces on one side the action of the marked (north) pole, and on the other of the unmarked pole of the magnet.

As it has been remarked, this is the fundamental fact, (the rotary movement accompanying the orbitual,) to which other facts of *electro*-magnetism are reducible. Magnetic bodies derive their power of attraction from the circular movement of force. Within them, extending beyond them to bodies of similar nature within certain distances, the polarity of magnetical bodies is owing to the secondary or rotative current which invariably accompanies what may be termed the primary motion of force.

Thus is the polarity of the needle upon the surface of the globe indicative of a current of force circulating around the globe in planes parallel to the magnetic equator, increasing in power from the equator to the poles, obeying the general law, increase of intensity with decrease of the orbit. M.

Ampere was the first who recognized the existence of this current, which has been imputed to the influence of the sun upon the tropical regions. That the current flows there is no doubt; the finger guide, the magnet, is authority; and it not only indicates the fact, to which we shall often refer, but, what is more important, points to a principle of universal application, — the onward primary movement of force is always accompanied by the secondary. Thus orbitual and rotary movement as the law of force is affirmed not only by the movements of the spheres, but by every rivulet of force which has strayed from the mighty channel of power, — to be predicated, not only of the career of worlds, but of the action of the most minute particles of matter.

The preceding remarks also indicate the action of force upon the magnet. By the touch of the loadstone there is given to the atoms of the needle the structure, which determines the flow of force through it in a fixed orbit. Force in any other direction gives motion to the needle on its pivot; force in the direction of this orbit flows through it, and it is unmoved by it. Thus it oscillates, a motion never produced by a stationary fixed force, but by the absorption and transfer of force. In one direction force gives motion to the needle; in another direction the force flows through it, presenting a perfect analogy to the oscillations of the pendulum. This idea will be explained as we proceed.

The attractive power of the magnet may be understood from the process of inducing magnetism. Dispose of two bars of steel in a parallelogram, connect them at the ends with two pieces of soft iron, thus making a circuit or orbit for force, and apply the loadstone; or in a horseshoe magnet connect the extremities by a piece of soft iron, making the circuit, and then apply the force through the medium of the loadstone. Thus you have polarity,—a direction of the circulatory force. The keeper, as the piece of soft iron is called, when attached, is a part of the orbit, and, while it is attached, the circuit of force being complete, the magnet has almost no attractive power. Remove it, and iron, a substance resembling steel in its structure, will be brought by force to finish out the broken orbit. If the loadstone be applied to a circular piece of steel, magnetism is as truly induced, but force revolves latent and unobserved. Such is the delicacy of the established orbit, that a fall upon the floor will often materially affect the power of the magnet; and magnets, to have their power preserved, should be laid away with the keeper, that is, with the completed orbit. Some additional facts will throw more light on this subject. The concussion of the steel by a hammer, causing it to vibrate strongly, will aid the process of inducing magnetism. Thus force is supplied, or the molecular movement enables the atoms to arrange themselves in the line for the orbit. After the orbit is established, a sudden jar may disarrange it, by excess of atomic motion; and oxygen, uniformly the destroyer of structure, always annihilates the orbit.

An unlettered man, on looking at the action of a magnet, exclaimed, Why, it appears as if it let out a loop of cord to draw something into its place. Well did he indicate the action of force in completing its orbit, and changing it from the elliptical to the circular. Many attempts have been made to convert the magnetical force into motive power for economical purposes. In vain; and the reason given for the failure is, the short distance at which magnetic force can be made to act. But a more general principle is the cause of the disappointment,—one which, when understood, will prevent any further attempt. The action is such as itself to complete the circuit, and the revolving interior force is locked up, giving cohesion to the mass. The completed orbit of the magnetic force is beautifully illustrated by pieces of fine wire, cut in lengths of about 1-8 inch, which may be suspended from the magnet, forming a chain or connected loop from one pole to the other. The soft iron receives polarity by this contact without the power of retaining it, except when the circulation is maintained by the permanent magnetic structure of the steel. We notice, too, the accumulation of intensity at the ends of the magnetic bar, the ellipticity of the orbit developing as ever the increased velocity of force, just as the comet is drawn with accelerated speed in the sharpest part of its elliptic course.

We notice in this place, as an idea growing out of the preceding, the difference of structure between the electric and the dielectric, the good and the bad conductors of force. In a strong metallic conductor the force condenses itself, increasing in intensity with the decrease of its orbit, — in other words, in the electric it establishes its orbit; while in the dielectric the orbits for circulation do not exist, or feebly exist. Says Daniell, "In the dielectric the force cannot travel from one end to the other to accumulate at the ends, like the electric forces in the insulated conductor." We refer to this fact now, as it throws light upon the action of imponderables, for instance, force destroying wood, while it enlarges iron; but mainly because this distinction will add much clearness to our conception of the important conditions of mechanical action in the pressure and friction of bodies in contact.

Whence is animal force? We are prepared in some degree at least to detect its source, if it lie not too deep in the mysteries of the living organization. The proximate cause of animal heat (force) is unquestionably respiration; but physiology has not yet arrived at a distinct enunciation of the principle, by which the change in the condition of the inhaled and exhaled air deposits force. It is presumed that it is the breaking up of animal tissues, which are exhaled as the base of carbonic acid. This slow combustion, it is said, evolves heat.

The accelerated action of the lungs by the need of force for unwonted exertion, and the continuous full play of the lungs as denoting permanent strength, tell us as plainly as facts can speak that force comes proportionally to the act of respiration. If our general view be correct, we shall find that the air has parted with force on its contact with the lungs. It is even so. The air expired is heavier than when inhaled. It has been deprived of force. The formation of carbonic acid is not only the principle of heat, but the condition of the force of life in all organic structure. The reason of its transfer of force to the blood, and through it to all the separate parts of the system, is only to be known by understanding the nature of life, which is perhaps wrapped in obscurity by the very belief that its mysteries are inscrutable. But should we ever admit that any thing which bears the stamp of the finite is unintelligible?

Animal force develops itself under the conditions of heat, light, electricity, galvanism, and, it is said, of magnetism. It is circular in its flow through the system; the heart and brain establish its polarity or direction. Electricity and galvanism will move the muscles, taking the place of the muscular force.

Do we wish for more proof of the identity of heat, current affinity, galvanism, muscular force, of all and every form of power? If you elevate a body to make it rotate at a higher level from the earth, thereby increasing the circumference of the circle it must describe, will not either and all of them give to the body elevated its required force? Are not each and all of them resolvable into the force which moves the spheres?

We have passed over this examination of the streams of force with a hurried glance. If these excite our wonder and admiration, what is the feeling with which we fix our attention on the fountain itself, from which these come as the accidental spray and tossing of the ocean ever spread out before us, — the force which moves the system of systems, the creation of an eternal and almighty God, revolving in circle upon circle through limitless space. We stand speechless at the thought of this boundless power upholding the heavens and the earths; yet boundless as it is, the strength of creation softly and kindly ministering to the breath of the sleeping child!

Whewell expresses the idea of the identity of force in the one hundred and third aphorism of the Philosophy of the Inductive Sciences, in these words: "Mechanical, chemical, and vital forces form an ascending progression, each including the preceding. Chemical affinity includes in its nature mechanical force, and may be often practically resolved into mechanical force," &c. Chemists generally indicate the identity of force by their communication of the fact, that the same matter in its atomic divisions has the same affinity for gravitation, for heat, and for electricity. Thus is pointed out, analytically as well as synthetically, the identity of force. We need not enlarge, as every chemical manual gives the results, and it would be the mere task of the copyist, transferring the sentences of the common books on the science of chemistry.

"We find," says Humboldt, "among the most savage nations, as my own travels enable me to attest, a certain vague, terror-stricken sense of the all-powerful unity of natural forces, and of the existence of an invisible essence manifested in these forces, — whether in unfolding the flower and maturing the fruit, in upheaving the soil of the forest, or in rending the clouds with the force of the storm. We may here trace the revelation of a bond of union, linking together the visible world and that higher spiritual world which escapes the grasp of the senses. The two become unconsciously blended together, developing as a simple product of ideal conception the first germ of a philosophy of nature."

The great error of science is its too minute classification of facts, and the attempt to find for each class of fact a distinct and definite law. Facts are never isolated; there are, in philosophy, no lines of demarcation between them. It is the poverty of man's intellect that thus subdivides. Take, for instance, the classes, departments, orders, kingdoms of organic structure. How they fade away one into the other! How clear on this point is Agassiz, in his Philosophy of Natural History! Other strong minds follow his investigations. He does not look upon organic structure only as a succession of classes; he does not affirm laws of local or of special application alone. He views life as a whole, and sees the general law which gives a oneness to creation. He combines when others analyze; he looks for simplicity where others see diversity. How true is the law of the mind to the law of nature! The pure conception of the savage seizes the idea of the unity of nature, and the truly great philosopher offers the same as the deduction of the reasoning power. It is in an intermediate order of minds that we find the errors, the accumulation of unassorted facts, the confusion of conflicting theories. The first thought, natural to the mind, is unity, — the last and highest conceptions return to the thought of the pure and simple reason.

CHAPTER III.

"PHYSICS DISCOVERS CAUSES FROM THEIR EFFECTS AND EFFECTS FROM THEIR CAUSES." —*St. Victor.*

CURSORILY must we pass through our examination of capillary attraction, heterogeneous and homogeneous attraction and repulsion, and chemical affinity; cursorily, because we need not devote attention to the minor phenomena, as the consideration of the more prominent involves that of the lesser; for general laws grasp the points, however minute, as well as embrace the aggregate of points which constitute the universe, force under general laws determining the place and position of all things, the minute as well as the extended.

It is sympathetic motion which forms the continuity of worlds; and if a mass be formed, it is also by sympathy of the motion of its parts,—it exists as a mass by this sympathy of motion,—if it be rent asunder, it is by jarring and conflicting motion. If there were not harmony of motion among the spheres, the solar system could not exist. If there were not sympathy of motion among atoms, masses could not exist. From this latter idea are deducible homogeneous and heterogeneous attraction, the phenomena of filtration, of exosmose and endosmose, and generally the attraction and repulsion of bodies at distances not too remote for their reciprocal action on each other through the connecting media of air, water, or other substances.

Capillary attraction is certainly inexplicable on the ground of the attraction of matter. As generally understood, matter attracts in proportion to its bulk. But water, attracted by the thin glass tube, escapes from the attraction of the earth which retains its hold upon the glass tube, yet lets the flowing water elude its grasp, and run up in perfect independence. But the centre of the glass tube has no more attracting power than the ends; and, if the glass be filled with water, the tube ought to attract downward as well as upward. Indeed there is in this phenomenon not one solitary fact to show that the glass attracts the water, except that the water in the tube is concave, rising where in contact with the glass; nor does this manifest the attraction of the glass, but an attraction upward. Surely if the sides of the glass attracted, the area under the surface would attract downward as well as the area above attract upward. Perhaps wetting the bore of the glass takes away the power of attraction!

It has been remarked that "the boldest imagination can hardly form a conception of the undulations rendered latent without annihilation, laid up in store as it were, and capable of being drawn forth in full measure and intensity." This appears to be the idea which leads to the understanding of the phenomena of capillary attraction. These undulations, latent in the water, are brought into the consentaneous motion of the mass inclosed in the narrow tube. It appears as if the range of molecular action were reduced by the confinement, and the no longer used atomic force is resolved into rotative force at a higher level. In mercury, the action is perhaps the converse; the atomic force of the glass may be increased, and force be withdrawn from the fluid metal. We thus offer a more consistent theory than the old theory, even if it be not true, and its truth may be rendered more apparent by our proposed examination of the properties of fluids.

We read, however, that "it has been proved by experiment that mercury will rise above its level in the same manner as water, when both the tube and mercury are perfectly dry. The mercury being dried by repeated boiling, is freed from its humidity. It appears, therefore, that the depression of the mercury below the level is occasioned by a little moist film, which attaches itself to the interior surface of the glass, and thus by its interposition, perceptibly weakens the attractive force of the glass," &c. The effect on the phenomena of capillary attraction will be further elucidated in the prosecution of this inquiry. One thing is certain, the rise of water is not from the attraction of the glass; for the glass would attract downward as well as upward, especially as aided by the attraction of the earth. It has been supposed that the ring of glass above the surface is the attractive cause. By others, as by Dr. Hamilton, "that the fluid is supported by the attraction of the annulus contiguous to the bottom of the tube." The bulk of supported fluid is not proportional to the surface of the glass with which the fluid is in contact. Of course, "gravitation" in capillary attraction is out of the question.

Elasticity is the property of bodies, the particles of which have a wider space and larger vibrations, which vibrations can be lessened for the time by the application of outward force. Glass, for instance, is extremely elastic, and the wide separation of its atoms gives it, probably, its transparency, and perhaps its peculiar properties in relation to electricity. Air is highly elastic, and is distinguished by the readiness with which it receives and transports the vibration of denser media of force. It acts as the conveyer of motion, as the distributor of power, trembling before the least impulse, yet spreading out the vibration in every direction, till the moving force is equally distributed and the equilibrium is again restored.

Cohesion is supposed to be a branch of the attractive force of matter. This force, it is said, "connects the particles of bodies together in the solid form with greater or less energy." To increase or diminish this force so as to keep the particles from actual contact in adamantine hardness, and to give the different energies of cohesive power, an antagonist power is imagined, the repulsive force of matter, "separating the particles of bodies from each other with greater or less energy." Like Mahomet's coffin, the atom hangs midway in the rest of perfect equilibrium. But, alas! bodies *in equilibrio* move at the softest touch, and the particles of the solid iron would tremblingly oscillate as the slightest breath of wind passed over them.

A better knowledge of the action of current affinity, that is, the acknowledgment of the laws of force developed by the recent experiments of distinguished chemists, has led to the supposition that cohesion is induced by the revolution of electrical force through the mass, though the belief is not yet generally or distinctly avowed. "Schelling, in 1803, says,"—we quote from Whewell,—"magnetism is the universal act of investing multiplicity with unity." Accordingly we find Schelling welcoming, with a due sense of their importance, the discoveries of Farraday: "When he heard of the experiment in which electricity was produced from common magnetism," he thus reasoned;—"we have three effects of polar forces,—electro-chemical decomposition, electrical action, magnetism. Volta and Davy had confirmed experimentally the identity of the two former agencies; Œrsted showed that a closed Voltaic circuit acquired magnetic properties; but in order to exhibit the identity of electric and magnetic action, it was requisite that electric force should be extricated from magnetic. This great step, he remarked, Farraday had made in producing the electric spark by means of the magnet." Whewell continues,—"Although conjectures and assertions of the kind thus put forth involve a persuasion of the pervading influence and connection of polarities, which persuasion has already been confirmed in many instances, they involve this principle in a manner so vague and ambiguous, that it can rarely in such a form be of any use or value." From the opinion of one so distinguished by reasoning power, it is perhaps presumption to differ; but it is deemed that polarity, or in other words, circulating bands of force are the only possible tie for continuity of mass, for the connecting bonds of the solar system, for,—in the words of Schelling,—"the act of investing multiplicity with unity." We will give our reasons for this opinion as they have occurred to us.

There has been established a connection between magnetic, electrical, and optical, and between these and crystalline polarity; and crystallization is the taking up of the bond of the cohesion of solids. Bodies never crystallize, except when their elements combine chemically, and solid bodies which combine, when they do it most completely and exactly, also crystallize. The forces which hold together the elements of a crystal of alum are the same forces which make it a crystal. There is no distinguishing between the two sets of forces. Accordingly, Berzelius asserts that the regular forms of bodies suppose a polarity, which can be no other than the electrical or magnetic polarity. Again; "the identity of a crystalline and optical polarity is too obvious to need insisting on." Thus much from authority of these latter days; and we quote the words of Newton: "Would it were permitted us to deduce the other phenomena of nature from mechanical principles, and by the same kind of reasoning; for many things lead me to suspect that all these phenomena depend on certain forces by which the particles of bodies are either urged toward each other, through causes not yet known, and cohere according to regular figures, or are repelled and recede from each other; which forces being unknown, philosophers have hitherto made their attempts upon nature in vain."

We are disposed to believe that the ultimate particles of bodies have an orbital or rotary motion, intense in proportion to the circumscribed space in which they move. In every mass, also, and at times extending beyond the mass, (as especially the object of perception in the magnet,) this motion exists, the phenomena of which are those of polarization, and the result of which is cohesion together with the phenomena of friction, and perhaps of chemical affinity. The force revolving in the mass is cohesion, out of the mass friction or pressure, and the change of the structure of the mass is chemical action. These ideas we will attempt to develop.

The magnet, which by its polarity indicates the passage of the current of force, by its attractive power will indicate the nature of cohesion. The magnetized steel has a circulating force, which will leap out of the boundaries of the mass to bring to its embrace another mass of iron or steel. Close the circuit with the keeper, and the force revolves within, as the bond of cohesion between the keeper and the magnet, and not only so, but as the bond of cohesion between the atoms of the magnet. Thus every mass has its circulating bond of force, beyond which its particles may not stray; and it would seem as if nature, having no fondness for mystery, hermetically seals up none of her processes, but protrudes the force of cohesion of the steel, that it may be perceived, and that what preserves the continuity of every mass may thus be known. In the horseshoe magnet with its keeper the current of force circulates, latent and unperceived but by the pendent keeper which completes the orbit. In the circular steel the force is just as active, without indication except by the general cohesion of the mass. So in the bar of iron. Thus by the simple law of nature there is continuity of mass, oneness from multiplicity; the bond is the revolving force. Its precise manner of action may be traced out.

Science as yet has hardly recognized the existence of electricity and magnetism, except in the incidental exhibitions of their power in the magnetic needle, in the flash of the lightning, in the artificial spark rubbed from the glass cylinder in the lecture-room. But they are everywhere a mighty branch of the great stream of force upholding the universe, connecting together the creation of God.

To repeat a former illustration, would you ask the nature of the cement of the brick building, which, when it was torn down, filled the air with the dust of lime? Break up the cohesion of the mass, and heat is evolved; — can you then separate heat in your mind as something opposite to, and distinct from the electrical force? Rub substances together, electricity starts forth; — does not electricity then circulate to the very surface of the mass, and often beyond the mass?

Pressure is therefore the connection of foreign bodies in near contact, by the force of cohesion extending itself from mass to mass. The phenomena of friction abundantly prove this. First, because, though friction bears a general ratio to the quantity of matter rubbed together, it is not an invariable quantity, as it would be if it depended solely on the weight of the mass. But the subject must be examined in its details.

Friction, or the degree of friction, if it were induced by weight or gravitation, would be almost, if not altogether, annihilated in a smooth body moving horizontally on another smooth body, which is not the case. If the degree of friction were caused by weight, it would be a constant value, measurable by the weight, which is not the case. If the degree of friction depended on smoothness of surface, it would be measured by the polish, which is not the case. If the phenomena of friction arise mainly from the passage of electricity between the rubbing bodies, it will have a general relation to the extent of the masses; it will be variable in degree; it will be increased or diminished according to the electric nature of the substances; it will be increased by the presence between the rubbing substances of conducting bodies, and decreased by the presence of non-conducting bodies; it will be increased by the rubbing together those substances inducing magnetical or electrical action, and the converse, — all which are settled facts.

Thus M. Colomb has determined by experiment that friction is increased between bodies of the same weight and the same smoothness, by the act of rubbing together, — by the gradual induction of the electrical force. Bodies of different structure obtain their maximum of pressure in different times. For one instance, a weight of one thousand six hundred and fifty pounds in contact was moved at first with a force measured by sixty-four pounds; in the lapse of three seconds, it required a force of one hundred pounds; in three days, six hundred and twenty-five pounds would hardly move it.

Again; to create the electrical state it is not necessary that the bodies should be rubbed together, it being induced at times by contact merely. The iron wheel of the locomotive only touches the track, bears upon it, and is lifted. If there were no magnetical action, the wheel would turn at the axle, but not impel the train. Yet the circumference of the wheel must adhere, for propulsion. If the non-conducting substance, ice or snow, is on the track, then this electrical action is so much reduced that the pressure is not sufficient to turn the wheel; or rather the wheel turns by the conducting rod, but for the want of electrical action, pressure, friction, weight,—call it what you will,—it takes no hold upon the track, and the locomotive cannot advance. A practical engineer in conversation indicated his knowledge of the fact, that the passage of the train over the track induces magnetical action. Pointing to a pile of unused railroad bars, he said, "Those bars have been exposed to the air a few days only, and they are red with rust, while the bars laid down for years in constant use scarcely show rust. Is it not," asked he, "the effect of magnetism?" When we ride swiftly over the winter railroad,—the snow,—one is apt to think of the smoothness of the snow only as decreasing the friction of the runners of the sleigh, forgetting that ice is a non-conductor of that which determines friction.

Of course, when we speak of friction, we know that it is modified by smoothness of surface,—that rough surfaces cannot be rubbed together without force to break up the cohesion of the projecting points; but, aside from this with equal polish, the extent of pressure is measured by the passage between the surfaces of the electric or magnetic force. Thus of iron rubbing on iron the pressure may be indicated by the fraction $\frac{1}{6}$, while iron rubbing on copper, with no smoother surface, stands at $\frac{1}{35}$; for the reason that iron and iron excite the more intense magnetical action.

The effect of oil as a non-conductor, decreasing pressure and friction, is perfectly indicative of what constitutes pressure and friction; and so is the converse practical act: if we wish to walk securely on ice, we must excite electrical action, put on the woollen moccasins; "coarse woollen and glass or ice excite the electric action." Thus empirically do mechanics excite or repress friction, often without the least reference to principle. Oil and ice are smooth, slippery, &c. Why slippery, has not been indicated. In the manufacturing of plate glass, two plates placed with their surfaces in contact, are sometimes found to adhere with such strength of cohesion at the former surface of separation, that, in the act of separating them, the split is not coincident with the line of junction. Oil, or any other non-conducting substance, would prevent this formation of cohesion between them. We have the following facts from one practically well acquainted with machinery. A piece of smooth cast iron revolving on other cast iron, without oil, will often file away the iron on which it revolves with great rapidity, the iron dust falling as saw-dust from the sawed wood. A revolving spindle, or other wrought iron body, moving on a wrought iron axis, without oil, will often increase its friction until the motion is stopped by it, and the two pieces of iron will be found cemented or welded together, so as not to come apart at the former surfaces, and when separated the grain of the iron is broken. "I have been told," said he, "of mills being stopped by this adhesion of the gudgeon of the propelling wheel for the want of oil." Other instances will occur to every one, such as the close adhesion of two pieces of glass when wet, their slight adhesion when greased; the two drops of water kept from union by a thin coating of dust; the stratum of oil over water preventing the water from freezing, though cooled below the point of congelation; air prevented from penetrating the wire gauze when wet; the non-ascent of water in capillary tubes composed of a material that cannot be wetted; the spreading of oil in woollen cloth; the repulsion, as it were, of oil and water; the rise of water in the pores of a lump of sugar; the process of the solution of the sugar. In fine, the whole class of phenomena of what is called homogeneous and heterogeneous attraction, is traceable to this principle, the extending, imparting, diminishing, taking away the bonds of cohesion by the presence of body with body, according to their electric and dielectric natures.

We read in old philosophy, "that flat and smooth surfaces of metals, glass, &c. do also adhere with considerable force, but with some other bodies a certain artifice is required for the purpose, namely, the interposition of some fluid, as oil or water, &c., or of some substance that may be applied in a soft state, which will after coagulate and grow solid, as tallow, wax, and fluid metals. Where something is interposed, the cementing or adhesion seems not to take place between the two surfaces, but between each of those surfaces and the interposed substance; for, in the first place, it seems strange that the surfaces *should have a greater attraction* when something is interposed than otherwise; and secondly, it has been found that the adhesion differs according to the different substance inserted between."

Hence adhesion is not the attraction of the adhering substances. The artifice of cements is the interposition of some substance, the electrical nature of which is such as to permit the force of cohesion of the two masses to pass and circulate through both ; it is the enlargement of the orbit of the circulating force ; it is making common orbits from special orbits ; it is extending the bond of continuity. The act of soldering is an instance of this process most perfectly accomplished, as the interposed substance is of the same structure and elementary constitution with those united. We notice also the converse in the amalgams which mercury forms with other metals ; the mercury becoming to a degree solidified, the other metal to a degree made fluid, so as to produce the mean of the cohesive force of the mingled metals.

There is also a modification of the force of cohesion by the mechanical addition of force. Let a heavy weight be suspended by a copper wire, the wire will remain of the same length ; but by striking the suspended weight so as to give a vibratory motion to the particles of copper composing the wire, the wire will become longer ; — there is added force, greater molecular motion, wider orbits, and of course a lengthened wire. Crystals form in a saturated solution by the addition of force in agitating the water. Crystals also are formed on foreign substances, by the extension of the force of cohesion from the bodies on which they form. Light, which is a form of force, induces crystallization.

When one substance is rubbed upon another, the body in motion preserves its own force of cohesion, and breaks up the cohesion of the body at rest on which it moves. This fact is illustrated by the tallow candle fired from the gun, the tallow by its velocity or intense present force preserving its own cohesion, and breaking up the cohesion of a far harder body; — also by the action of the comparatively softer grindstone, in its revolution wearing away the harder steel; and by the diluvial striæ, which are grooves worn into the surface of the rock, by the rapid passage of an angle of a rock no harder than itself.

The destruction of cohesion by the act of cutting with a keen instrument, is much more nearly connected with electrical action than is generally supposed. It is well known that the electric spark will, under certain conditions, perforate a glass jar; its perforation of the pasteboard card is a frequent experiment. Steel has high magnetic properties; and the intensity of the action of electrical force is increased in proportion to the reduction of its orbit of circulation. We have, then, in the sharp edge, for instance, of the razor, the condition of great electrical activity. It has often been remarked that the keenness of the edge varies without assignable cause. A condemned razor, used accidentally, will sometimes be found to have its edge restored by lapse of time. The strapping of the razor is usually on non-conducting substances; but the edge is often restored by strapping on metallic surfaces. Between both a fine edge can be obtained, as if the keenness depended on one exact degree of magnetical induction. Strapping does not wear away the steel, or make the edge thinner; for strapping on the palm of the hand will often be effectual. The nature of the stroke which severs, strengthens the opinion that the act of cutting with the keen edge is not dependent on force to break cohesion, but on the electrical induction. Thus by establishing correct theories in relation to actions of an enlarged sphere, even trivial operations and the common arts may be improved.

From all the foregoing we are led to believe that the

circulation of force is the bond of cohesion, and that the acts of separating and of connecting, of lubrication and of cement, of pressure and of friction, are electrical phenomena.

To Boerhaave is assigned the honor of originating the idea that chemical affinity is a peculiar force, having under its control the elementary particles of matter. By affinity in this use of the word is meant attraction. The word affinity was first used in France, to avoid the word attraction, "which had the taint of Newtonism." Elective affinity expresses a choice of matter in the exercise of its attractive power. Berthollet asserted that affinity is not elective,— that, when various elements are brought together, their combinations do not depend on the kind of element, but upon the quantity of each which is present, that which is the most abundant entering most largely into the resulting compounds. This doctrine was assailed, and the celebrated chemist Berzelius says, "Berthollet defended himself with great acuteness, which makes the reader hesitate, but the great mass of facts finally decided the point against him." The theory of Dalton, at present the accepted theory, rests upon the idea of substance as well as upon that of choice by the attracting power.

If we affix the usual meaning to the words choice, preference, election, the idea is very unphilosophical. If the words be retained in chemical science, they must be understood as conveying no other meaning, than that chemical action is determined by the laws of force, which destroy and rebuild the structure of masses. The laws of force never vary; their seeming changes in relation to the movements of particles come from defective observation, and reason must supply the deficiency of perception. We must interpret the minute by knowledge gained from the extended. Thus, water and oil will not chemically unite. Why? The action of oil in preventing the union of iron with iron is from its dielectric properties. It is an electrical phenomenon. So water and oil, from their opposite electrical natures, from the want of force equally and commonly diffused, will not unite chemically. We cannot form one mass or volume from the two. Add an alkali, and the union of the three substances is formed. What analogical action is there in the extended? Does not non-conducting matter separate bodies or repress friction? Is not the action of cements founded on the relations of the varying capacities for force? We shall again advert to the subject, perhaps better prepared to entertain the position, that the law which determines union is the capacity of the bodies present equally to diffuse the force of cohesion, and the more intimate bond of connection. Chemical union is of the same nature, the one mingling the elements, the other mingling the particles; the one determining the chemical nature, the other determining the mass of aggregation; the one effected by the force of cohesion, the other by the force of constitution. Both are of the same nature, and under the action of the same laws.

It may be objected to this view that the generalization is too far extended. But we can hardly err on this side. The present state of philosophy seems to demand generalization, — the grouping together of the heretofore isolated action of nature. Nature has been too much parcelled out, — too much separated into artificial departments. There have been too many tribunals set up, each with its own code of laws. Such divisions are not characteristic of the simplicity of truth; but have been made from the necessity of the case. There is an error in the theory which relates to general principles. Gravitation cannot be brought down to the particulars; it will not apply to the minute. It has to be added to, or taken from; it has to be modified for every class of phenomena. Thus have we gravitation, attraction, elective attraction, capillary attraction, attraction of cohesion, attraction dynamic, attraction statical, attraction between elements of one kind, attraction between masses of different kinds, and so on, almost without end; and in place of science giving method to the mind, clearness and distinctness to thought, the intellect is embarrassed in its attempts to assign to nature the modicum of order which the theory itself may have in minds of the greatest strength.

CHAPTER IV.

"WE ARE NOT TO ADMIT OTHER CAUSES OF NATURAL THINGS THAN SUCH AS ARE BOTH TRUE, AND SUFFICE FOR THE EXPLANATION OF THEIR PHENOMENA." — *Newton.*

THE third law of motion is usually expressed in these words: "Action and reaction are equal, and in contrary directions; that is, equal and contrary changes are produced on bodies which mutually act on each other."

Sir Isaac Newton seemed to consider the equality of action and reaction in the light of an axiom deduced from the relation of ideas, rather than an idea induced from the observed relations of things, — to be believed by a law of the mind rather than from observation of phenomena. From this law of the mind, "he presumed, because the sun attracted the planets, those also attracted the sun; and he is at much pains to point out to astronomers, the phenomena by which this may be proved when the art of observation shall be sufficiently perfected." But Mr. Robison, in his Elements of Mechanical Philosophy, considers this doubtful, and says that, because a magnet causes the iron to approach towards it, it does not appear that from a law of the mind we are led to the belief that iron also attracts the magnet. He therefore states it as a fact with respect to all bodies "on which we can make experiment or observation *fit* for deciding the question." Maclaurin does not consider it as deducible *a priori* from abstract considerations, but thinks of it as the effect of an arbitrary arrangement. Mr. Stewart says that it is "more safe and more logical to consider it as an experimental truth, without venturing to decide on either side of the question." The origin of the idea we trace, however, neither to the relations of ideas, nor to the relations of things, but to the action of the human body in the exertion of its muscular power. When we attempt to move a heavy mass, there is a feeling as if it withstood the strength, reacting as it were against the effort put forth. This is probably the origin of the idea, and it serves in almost every treatise on the subject as an illustration of the law.

The observed facts which give support to the law are all such as manifest the continued action of the same force, with its induced motion changed in direction;—as in the impingement of billiard balls;—in the resistance of water to the motion of a ship;—in the counterpoise of two heavy bodies, where one descends with the same quantity of motion with which the other ascends;—in the vibration of the pendulum, the rise being equal to the fall;—in the rebound of a ball in proportion to the strength of the impulse given. The law, therefore, as it stands, affirms with some obscurity the doctrine of the independent existence of force, of its transfer to other bodies in whole or in part, and of the change of direction often induced if it continue in the same body. In this point of view the law stands, as has been said, as the basis of mechanical philosophy, and rightly understood expresses a universal law which applies to all motion in the universe;—Action proceeds from force, reaction proceeds from the same force. The result of any degree of force is continued action. It may be divided, giving two bodies equal and contrary action, or it may continue in the same body acting in a contrary direction. The action of force is uniform; its transfer and line of direction contingent.

There is an unintelligibility in the doctrine of reaction as usually explained. It is very difficult of statement in conformity with the present theories of motion and the causes of motion. We turn to a philosophical treatise of rather ancient date, and find these instances given: "When a man strikes one hand against the other, the blow is felt equally by both hands; if you strike a glass bottle with a steel hammer, or the hammer with the bottle, in either case the bottle will be broken if the blow be sufficient; if a stone be tied to a horse by means of a rope, the horse in dragging the stone will exert a degree of force equal to the resistance of the stone, for the rope will equally pull the horse toward the stone, and the stone toward the horse." But then, the error was too palpable, and the writer destroys the law at once, adding, "and in fact the stone will not follow the horse, unless the power of the horse be greater than the power of the stone." Indeed, there could be no motion under the law as usually stated, for all action would be met with equal resistance. It is however sometimes stated in such a manner as to give the idea that force is increased by being communicated. Dr. Arnott, in his Elements of Physics, says, "When any elastic body, as a billiard ball, strikes another larger than itself and rebounds, it gives to that other, not only all the motion which it originally possessed, but an additional quantity, equal to that with which it recoils, owing to the equal action in both directions of the repulsion or spring which causes the recoil. When the difference of size between the bodies is very great, the returning velocity of the smaller is nearly as great as its advancing motion was, and it gives a momentum to the body struck, nearly double of what it originally itself possessed. This phenomenon constitutes the paradoxical case of an effect being greater than its cause." These statements must always be understood in conformity with the principle, that every degree of force gives its definite amount of motion, though the quantity of motion may appear different from acting on different quantities of matter.

We will, however, state the doctrine of resistance in the words of Prof. Daniell, who is distinguished for accuracy in the enunciation of his opinions. He says, "In lifting a lump of iron or lead we are conscious of opposition, and consequent exertion." Of course; for the application of strength animal force is always attended by effort, by exertion; the effort is not because the iron or lead resists, but because to raise it requires the application of force for its rotation at the higher level. He continues,—"In drawing a bow, we feel an opposing resistance which we denominate elasticity, a gradually increasing opposition which our utmost force may not be able to overcome." . . . "If the opposition be suddenly withdrawn and the bow be allowed to act, on its return the arrow will have a projectile force communicated to it." All this is true. The muscles of the arm by force separate the atoms composing the one side of the bow, and press together the atoms on the other side of the bow. When the action of the muscles ceases, the atoms of the bow return to their normal position, and the force that disturbed them is communicated to the arrow, which rushes forward, gradually imparting its force to the air, till exhausted it falls to the ground. Here there is no reaction whatever. But we find in this instance a beautiful analogy between the action of the bow and of the bent arm, in the will to straighten the arm, and in the will of nature to straighten the bow.

The idea of reaction, as we have said, undoubtedly came from the idea of resistance in the exertion of muscular force. We cannot with the hand move a body without feeling what is called reaction; and probably the best instance that can be given is the result of the action of standing in one boat and pushing off another boat, in which case resistance is felt, and the reaction gives the contrary equal backward motion. But it is most readily explained. The action of the body in this case is like that of a bent spring with one end in each boat, and of course it acts equally on both in opposite directions.

The author just referred to asks this question, the true solution of which more fully explains the action of the spring, though his answer is obscure: "Is a spring stretched by the sum of the forces of the two arms, or by one only?" This is his reply: "Let one end of the spring be attached to a fixed support, and it will be found that the same effort of one arm, or one weight, will stretch it to the same amount as before; the force of the other arm or other weight only measured the resistance, — the amount of reaction, which necessarily accompanies the action, and which is now borne by the support." We reply, there is no reaction whatever; the needed force, whether from weight, from one arm, or from two arms, is supplied. A given degree, a measurable, determinate quantity of force must be applied whenever and however the spring be bent; but this force must be either applied to both ends, or one end must be secured and all the force applied to the other end. These are the conditions on which the spring is bent, or on which the force is applied to the atoms of the spring; without these conditions, if force be applied it will move the spring upward, downward, or sidewise, as a whole. Here "resistance" is the condition on which the spring can be bent. The strings of the harp on the application of force can be moved without being strung upon the instrument, but their melody of vibration comes from their tension. The arm not "resisted" at the shoulder could move no weight.

It is said that the blow of a hammer on an anvil is accompanied by reaction. We answer, — A part of the force which gives the blow is not transferred, but relifts the hammer, and, if the blow be unskilfully given, will declare its presence by a sharp vibration in the hand which holds it; a part is transferred to the anvil which vibrates the anvil, and the force spreading in all directions vibrates the air with a ringing sound. Let, however, the blow of the hammer fall upon heated iron, then the hammer is not relifted, and the force is mainly transferred to the iron, a small portion only escaping, giving a heavy vibration to the air. A most striking illustration of the distinction between absorbed force and transferred force is this: — If a string be pulled with great force and it breaks, "reaction" will cause him who pulls it to fall, — the unused force returns; let him draw wire through a wire plate, expending the same force, — when the end slips through the plate, he will feel no reaction, no shock, but will stand firm, because the force is safely deposited among the particles of the extended iron. We remember repeating the experiment with wonder at the different results, — the very results which might have been affirmed without the test of experiment, had we understood the laws of force.

The motion of a steamboat is generally considered as depending on reaction. The paddles of the boat strike the water, by the reaction of which, it is said, the boat is propelled. In fact the common statements with regard to reaction and resistance have thrown obscurity over the subject. We will trace out the action of the wheel of the steamer, in the hope that we shall find the truth, and the truth in this one instance will cover the whole ground.

It is the force of the steam which by proper mechanical appliances causes the wheel to revolve. If the wheel turned in a vacuum, all the force of the steam would be lost for propulsion, exhausting itself in whirling the wheel; — if the paddles revolved against an immovable body, like a ledge of granite, all the force of the steam would give propulsion to the boat; — if the wheel revolve in water, a part of the force will give a backward motion to the water, and the residue will give propulsion to the boat. In neither case does reaction move the boat, but the force of the steam; for the solid granite stands firm without force or motion, and of the water, neither the portion which is unmoved, nor the portion which moves backward, has the least effect in communicating onward motion to the boat. It might as well be said that the water at the bows propels the vessel by its resistance, as that the water at the wheel does; or that the iron track of the railroad gives propulsion to the locomotive, as that the track of the wheel in the water moves the boat. The force which impels the boat is from the expansion of steam, the water only giving direction to the motion of this force.

The force of the steam as generated is expansive, acting in every direction; it could be let off safely through the pipes into the air, thus wasting its strength, or it could be made to act on any part of the vessel, breaking her into pieces. To be useful in propelling the boat, it must have one line of direction given to it. This direction is relative to the earth. It must, therefore, in propulsion, have reference to something external, otherwise the "two ends of the spring" would act in the ship, giving to its parts motion in opposite directions; to give to the boat a motion relative to the earth, one end must be acting on matter outside of the boat. The action of the wheel on immovable granite is the securing of one end of the spring, so that its whole force may act in giving motion to the boat with the other end; and, as we observed, if the wheel revolved against an immovable body, the whole force would be used for propulsion. Acting against water, so far as the water is moved, the force is lost; so far as it is unmoved, or resists being put in motion, the force of steam is of use for propulsion. A familiar illustration of this is, that in rowing a boat near the shore, if the oar touches the rocks, much more force adheres to the boat shooting her ahead, since none of the oarsman's strength is wasted in giving motion to the water.

If we place the hand in water, moving it slowly, a slight resistance will be felt, and the resistance will increase in proportion to the velocity of the motion of the hand. It is said that a ball of iron will strike the water more forcibly than a ball of cork of the same size, and be more resisted than the cork, though both move towards the water with the same velocity. Therefore it is inferred that reaction is in proportion to momentum, momentum being the product of the velocity by the weight.

It is not so. The resistance to a body impinging the water is in proportion, not to its momentum, but to the quantity of the water to be moved out of the way, and the velocity with which it is to be moved away. A short ship, say fifty feet in length, with a momentum represented by 100, sails five miles an hour, and a ship one hundred feet in length moves, with the same breeze and proportional canvas, with a momentum of 200. The resistance increasing with the momentum, with the same force in proportion to the mass, the long ship under the theory would sail only five miles an hour. Practically, however, the longer the ship driven by the same proportional force, the greater the velocity; for the short ship with the momentum 100, loses say 10 of her force in moving away the water from her bows, in other words, in opening her path. In the long ship, moved with the momentum 200, it still requires only 10 for this purpose, leaving 190 for propulsion, which is 95 for each of the two halves. The long ship has then $\frac{5}{95}$ more force for propulsion. The per centage of force needed to open the path is less in the long than in the short ship.

Thus resistance to the vessel is not measured by the "momentum" of the vessel, but by the quantity of water to be moved, and the speed with which it is moved away. There is, in fact, no resistance in the case whatever; but a part of the force of propulsion is necessarily used in removing the water from the track. There are other circumstances which modify this loss of force, for instance, the shape of the bows. If the shape be such that the water is elevated, there is a loss of force to raise this water above the level. The form of the ship should be such as to establish a current from the bows to the stern. The less the surface of the water is disturbed, the better. For this reason a body wholly submerged is less impeded; for the path opened is but a circulation of the water from one end to the other of the moving body. If the body plunged be suddenly moved under water, before the current is established, the water will rise in a wave at the surface; after the current is established by the moving body, the surface is at rest. When the wave was raised, force was supplied for its higher rotation.

It appears, therefore, that what is called the resistance of the water to the propelled vessel, comes not from reaction, but is the loss of force in establishing the current by which her path is opened.

In the preceding statements the word momentum has been employed in conformity to the usual language. As was before observed, momentum is the degree of force present in action upon the body. It is the energy or strength of the motion. The intensity of force is measurable by the velocity it imparts; but as it requires more or less force according to the weight of the body, to move a body with a given velocity, in ascertaining the present force of a body in motion, its weight is one of the elements of the calculation. Momentum is therefore the force which moves the body, — that which it will impart when the motion is arrested. It is simply force, and the use of the word momentum is not only unnecessary, but it leads to a confusion of ideas to employ two words to express the same thing.

If a bullet be discharged from a gun through a door left ajar, it will pass through the wood, giving no motion to the door. In gunnery, cannon balls, to do the most injury to a ship, should be fired with small charges of powder, that they may not pass through the ship, but transfer their motion to it. In these cases, reaction increases with the lessening of the force, that is, if the ball pass through the ship, it does not injure the ship so much as if the motion of the ball be arrested, and its force transferred to the work of destruction. The gun recoils proportionally to the force of the explosion, of course; for explosive mixtures act in all directions.

The doctrine of vis inertiæ asks a passing comment. It is inferred from the first law of motion. Bodies in motion continue in motion from the inertia of motion; bodies at rest continue at rest from the inertia of rest; that is, bodies which move do move, bodies at rest are at rest. This is all. The idea receives some coloring from two facts. First, the progressive motion of a mass is in proportion to the force impressed, and takes place after a lapse of time from the application of the force to the atoms of the body. There is needed time for the force to be diffused among the atoms by atomic motion, before consentaneous or progressive motion is induced. The other fact is, that in most cases force in moving bodies is gradually applied.

Thus a bullet slowly moving will enter the water; for it can, as it were, await the tardy resolution of the atoms of the water into progressive motion to give it place. But the bullet from the gun is deflected. Its velocity is so great that it cannot transfer its force; for it would take many instants for the particles of water, moving with the same velocity with the bullet, to travel through their atomic space, and to induce progressive motion of the column necessary to be moved for the passage of the bullet through the water.

We now come to the consideration of another principle, which is often referred to in the explanation of the mechanical powers. It is this: The loss of time is the gain of power; and, conversely, the gain of time is the loss of power. It is not so. Why, let us ask, are there in the text-books of mechanical philosophy assertions like this: "Time is an important element of force"? Time has no reference whatever to force, or to motion consequent upon the force applied. It can neither make nor destroy it. It cannot increase or diminish it. A body with the same present force moves on forever with an uniform motion. Time can never change the quantity, quality, or direction of motion. Time is a succession of events. It is itself measured by motion,—by the harmonic uniform motion of the spheres. We measure motion by time indirectly; its direct measure is the motion of the earth. Velocity is more conveniently applied to the artificial divisions of time than to the earth's motion. It is, perhaps, better for common purposes to say that a body moves so many miles an hour, than to say that it moves so many miles while the earth moves one twenty-fourth of a rotation.

But the doctrine like that of action and reaction, and of vis inertiæ, has a coloring or semblance of truth from the misconception of facts. In two ways time has an apparent connection with force, as if time increased force.

First; animal or muscular force is produced by the animal by a succession of efforts. It is generated gradually, and of course imparted gradually. By time the force is increased or added to. Time does not increase power; but the muscles can supply more force in five minutes than in one minute. This idea is fully illustrated by the inclined plane. A horse, in dragging up a weight, has more time to produce the required force; but when the load is elevated, he has imparted just that degree of force which is required to elevate the weight directly, at once, vertically. The mechanical power of the inclined plane enables the animal to do by successive efforts what he could not do at once. He cannot raise the weight at once; he can raise the weight in time. A man cannot raise by one effort a ton of iron ten feet, but in an hour he can furnish the necessary force. Yet, it is said in a work on mechanics, that it requires no force to move a body horizontally, full force to raise it vertically, and that the gain of power by the inclined plane is in simple proportion to the angle it makes with the horizon!

Secondly; another misconception is of the law of falling bodies. The force of the falling stone, it is commonly said, increases according to the square of the time of descent. The force increases not with time, but with every line of descent. The descending body has, proportionally to its change of level, an increase of spare force,—of force unused at its lower level of rotation. The greater the fall, the more the force.

Thus in the use of animal strength, and in the use of the force of the falling body there is, not an increase of force by time, but an addition to the force received in time. A man's power of labor is measured by time; the power of the waterfall is measured by the space of descent, or for convenience' sake, by the time it falls, as the space and time are in one fixed ratio.

So simple is the explanation of the philosophy of the mechanical powers, that it is surprising that the subject should ever have been deemed intricate, or that so much mathematical labor should have been wasted on it. To understand the lever,—and, that understood, all the mechanical powers are understood,—does not require one to read Archimedes' mathematical demonstration, nor to understand the laws of statics, dynamics, virtual velocities, equilibrium of forces, and centres of gravity. Let one idea be present to the mind, and there can be no misconception. The force applied is the force that produces the result; time will not increase or diminish it, it ever remains of one intensity, capable of producing only one degree of motion; no lapse of time, no mechanical skill can change its nature or degree.

Without reference to loss by friction, it may be transferred through pulleys, levers, wheels, over inclined planes, whenever needed, and when used at the end of the line there is only, and just that force which was at first applied. It will there produce the same quantity of motion, neither more nor less, that its direct use would have produced without the intervention of the machinery. If by the lever one pound raises two pounds, the two pounds will be moved half as far; if a man raise a ton by means of a pully, he has put forth just the strength that would be required to raise a ton to the same height without the pully. His gain has been the time given for him to produce and apply the strength. He could not lift a cord of wood at once, but log by log he can supply the needed force for piling it. If a smaller wheel turn a larger one more slowly, there is still the same degree of motion in the larger wheel that was in the smaller; if the larger turn the smaller, there is still the same motion in proportion to the quantity of matter. It requires one degree of force to give one quantity of motion; a given force can do neither more nor less. This is rigorously true, and, if it be understood, there is no mystery about the mechanical powers. If there be a gain of power, it is distinctly traceable to the gradual addition of force from animal strength, or from the increasing force of a descending body. The measurement of these by time has led to the confusion thrown over the subject by the doctrine, that there is a gain or loss of force by time. It will not be necessary to remark particularly on either class of the mechanical powers, as it is believed that this general statement covers the whole ground.

The next supposed force to be examined is the centrifugal force, or the tendency of bodies moving in curved lines to assume a rectilinear motion. It is not exactly a force, but a tendency of the moving mass not to move as it is moving. In the *Philosophy of the Inductive Sciences*, we find the following explanation of the character of this tendency. "The centrifugal force is not a distinct force in a strict sense, but only a certain result of the first law of motion, and is measured by the portion of centripetal force which counteracts it." "The projectile force is a hypothetical impulse, which may at some former period have caused the motion to begin, while the central force, gravitation, is an actual force which must act continuously, and during the whole time of motion, in order that the motion may go on in the curve."

If we had no doubt of the truth of the theories which relate to the motions of the spheres, the reading of such passages would suggest a doubt. There is an obscurity of language, not arising from the want of power of expression, for no man has more power than this author, but because there was no distinct idea to be conveyed. In the first place, centrifugal *force* is not a *force*, but a result; not being a *force*, it is still measured by another force. Then, the force which is not a force is counteracted, — of course it is destroyed, it no longer acts. This verbal criticism is made only for the argument, that doctrines feebly expressed by strong men are of doubtful soundness.

It is supposed that the moon, for instance, received originally an impulse of motion in a straight line, but that, being constantly acted upon by the attraction of the earth, the straight line is changed into the curve; that the attraction of the earth which would, unchecked, draw her to the earth, is opposed by the centrifugal force; and that the moon, under the influence of forces acting in opposite directions, is held in her orbitual course, the original impulse giving the motion.

But as the moon's orbit is elliptical, she is at times comparatively nearer to, and at times further from the earth. Of course, in her changing distance, the centripetal and centrifugal forces vary in intensity. The earth attracts less or more according to distance, and the counteracting force changes relatively; they preserve an equilibrium. It cannot be that the force of opposition can remain with an equal intensity, when the force of attraction is continually changing. They therefore change proportionally, or to the same degree, so that the moon is guided in her course by forces perfectly elastic.

Again; if the original impulse of the moon's motion were an impulse of rectilinear motion, and the attraction of the earth acts in a straight line to her centre, how from two forces acting at right angles is the curved line of motion produced? An original impulse in a straight line of direction, and another impulse in a straight line at a right angle with it, would ever and invariably produce a straight line of motion at the mean distance, half way between the two lines of impulse. If gravitation changed the moon's motion, the moon must necessarily be brought thereby nearer to the earth; if it act on the moon impelled in a straight line, the resulting motion would be a mean direction, lessening the distance between the attracting and attracted bodies. Even Newton, at an early period of his speculations, held the opinion that the result of attraction on the impelling force would be motion in a kind of spiral.

That the orbit of the moon could be formed by the action of the two forces, has been nominally proved from the case of the cannon ball, which under the joint force of projection and gravitation describes a curve. But the case is not an analogous one. The curve of the projectile is occasioned by the regular decrease of the force of impulse. Its onward force is gradually lost by the resistance of the atmosphere, and thus "gravitation" brings the path described into a curve. In order for the attraction of the earth to give a curve to the orbit of the moon, the impulse of the moon must be gradually lessened by some retarding medium, as the cannon ball is retarded by its passage through the air, whence would result a spiral curve gradually bringing the moon to the earth. But never can the impulse in two directions of the rectilinear force result in a curve. We know that the reverse is mathematically proved, and that the nature of the curve which would be produced is pointed out; but the mathematics by which this is done, is of a peculiar kind introduced with the theory of gravitation, being rendered necessary by this theory.

What gives the elliptical form of orbit,—for instance, of the comet, which, at one time widens its path in the broad expanse of almost limitless space, and then narrowing its orbit, almost brushes in its rapid flight the face of the sun? It is not force of motion; for this force acts according to the law of the area described in every part of the orbit. It is not attraction; for attraction merely holds its own, increasing or diminishing according to distance. Nor yet is it centrifugal force; for this follows, and is measured by, gravitation. It is not space; for there is no grooved channel or marked track through its vastness. A sphere, perfectly a sphere, each atom of which was moved with equal force, would necessarily describe a circle. May it not be that it is the shape of the revolving body; that, if the shape of this world, for instance, is given, the ellipsis of her orbit may be determined; and that, if we know the form of the comet, its eccentricity of path could be understood.

The idea of centrifugal force grows out of a trivial, and perhaps misunderstood fact, and is invariably illustrated by it. We allude, of course, to the action of the sling. From the hand proceeds the impulse of motion, the string is the gravitating power, or that which by its tension measures this power, and the tendency of the stone to fly off in a tangent represents the centrifugal force. It is unquestionably the tendency of the stone to escape, and, if the string were to break or be let slip, the stone would no longer continue to move in the same circle, but the impressed force would give it another direction of motion. The force of motion of the stone is *from* the centre. There is nothing to give the stone its circular movement except the string; that broken, the impressed force gives the stone a curved motion in another direction.

So far is this action from being analogous to the motion of the moon, that it is the very reverse. The stone from the sling receives force acting in a line of direction from the centre; the moon receives from the centre of her orbit an impulse toward the centre. It is not a fortunate illustration to compare the effect of an impulse away from the centre, to an impulse affecting the body and drawing it toward the centre.

Centrifugal force, too, is often illustrated by the necessity which a man finds in running round a small circle, of leaning inwards toward the centre, "to counteract the centrifugal force." But we will quote a contrary opinion: "Let any man move in a circular or elliptical line described or marked out to him, and he will find no tendency in himself either to the centre or from the centre. If he attempt the motion with great velocity, or if he do it carelessly and inattentively, he may go out of the line either from the centre or towards it; but this is to be ascribed not to the nature of the motion but to our infirmity, or perhaps to the animal form, which is more fitted for progressive motion in a right line than for any kind of curvilinear motion." Thus says Lord Monboddo, and proves that philosophers are not agreed among themselves concerning a fact, which is assumed as a "standard example" of their mechanical doctrines. However it may be with a man running round in a circle, we feel quite sure that a revolving sphere would not have to lean inward to prevent its falling outward, or to lean outward to prevent its falling inward.

The action of the governor of the steam engine has also been referred to, in order to illustrate centrifugal force. The two heavy balls are so suspended that when not revolving they hang motionless; endued with activity in proportion to the velocity of their revolution, they rise, and describe wider and wider revolutions. The reason of the rise of these balls is the force communicated to them, from which they take a higher level of rotation,—a beautiful illustration of the views we present. They could not thus rise from the action of centrifugal force, were there any centrifugal force; it impels, if it impel at all, from the centre.

How could the moon be held in her orbit, if centrifugal force acted in any other direction than the direction opposite to the force of gravitation, which is to counteract it?

The breaking of the too swiftly revolving wheel is another illustration of the centrifugal force, which keeps the planets from responding to the attraction of gravitation. But the application of force to turn the wheel is at the centre, as in the sling it is from within outward. The central part of the wheel as truly impels the circumference, as the crank or rod impels the centre. There is motion induced under the condition, that the cohesion of all the matter in motion will bear the force applied at the centre. If too much force be applied, the crank or the wheel may break. The greatest force required is at the circumference, while the force is applied at the point the furthest removed. Every practical mechanic knows that there is no danger of a wheel's breaking if the motive force be applied to the circumference. But nature does not so unskilfully apply her motive power. Her force for revolution or rotation is present with every atom. It is equally diffused through the whole mass, every particle retaining its proportional motive power according to the orbit to be described; each moves of itself, and from its own energy; one does not push, draw, or impel another. There is for this reason harmony of motion, all things moving according to the order of nature. Those who require a balance of forces, conflicting powers, — those who compare the motion of the spheres to a school-boy's sling, or to the dizzying whirl of a man running in a circle, seem to degrade the mechănism of the heavens below the results of human ingenuity.

We pass from centrifugal force to its opposite, centripetal or attractive force, — gravitation.

It may be well to give an outline of the theory as presented by those who have the most unshaken confidence in its truth. The theory of universal gravitation asserts, that the force, by which the different planets are attracted to the sun, is in the inverse proportion of the squares of their distances, and that the force, by which the same planet is attracted to the sun in different parts of its orbit, is also in the inverse proportion of the squares of its distances;

that the earth also exerts a similar force on the moon, and that this force is identical with the force of gravity; that bodies thus act on other bodies besides those which revolve round them; that thus the sun exerts such a force on the moon, and that the planets exert such forces on one another; further, that this force, thus generally exerted by the entire masses of the sun, earth, and planets, results from the attraction of each particle of these masses, which attraction follows the above law, and belongs to all matter alike; also, that this attraction, operating on the "hypothetical impulse of motion," gives its spheroidal form to the earth, and their elliptical form to the orbits of the heavenly bodies.

This theory has been a settled doctrine of philosophy for nearly two centuries, and generally, if not universally, it is deemed an established truth. It seems to have passed away from examination, to be laid aside as that which need never again be the subject of thought. It is spoken of as ascertained; and its discovery and extended application are considered as among the most glorious achievements of the human intellect.

Thus it will be almost impossible, even if the theory be not founded in truth, to bring the minds of men to its re-examination. It has become associated with the universe from its most enlarged to its most attenuated parts. The attempt to substitute other theories, to assign other laws of force, may be deemed the wandering of an erratic mind, or the presumption of an ignorant man. But this research was commenced without reference to the public, and the results are presented, not presumptuously, but firmly, — not with the belief that we have fully developed the truth, that all our views are sound, that no mistakes are made, but with the conviction that our speculations are founded upon principles that fully deserve attention and examination. We appeal to those who believe that there is yet hope of progress in philosophy.

Attraction, as coming under human observation, is mainly believed on considerations derived from the fall of bodies to the earth, and from their supposed weight or pressure when at rest. This is the foundation of the theory. All else is assumption or deduction. The mutual attraction of different masses at the surface of the earth is supposed to corroborate the theory; but from the utter impossibility of measuring this slight action, or even of ascertaining if there be any, the mutual attraction of the minor masses being overwhelmed by the immense comparative mass of the earth, it is a proof about as conclusive as that the sun shines because a fixed star twinkles. The doctrine of attraction, therefore, rests on the observed fact that bodies unsupported fall to the earth.

If, therefore, the descent of a body can be fully accounted for by the spare force of rotation at a lower level, and its weight or pressure by the fact that it requires force to lift it up to a higher level, the reason on which the theory of gravitation stands is taken away.

The law of gravitation is not needed for the motion of the heavenly bodies. There is more truth to nature, there is more simplicity and beauty in the idea that the force of the revolving body is within itself; that its curvilinear motion is its natural motion; that it goes round in its orbit without needing the guidance and direction of central and tangential forces; that it can be trusted to the unerring energy imparted to it from the beginning. It needs no great presumption thus to affirm; for by ancient philosophers, and in more recent times by Copernicus, Galileo, Kepler, Des

Cartes, it was believed that circular motion was the natural motion.

Nor is the general acceptation of a theory any proof that it is the most true and most perfect theory that will ever be presented. How long endured the Ptolemaic system of astronomy, how firm was the belief of men in its truth! Those who first questioned it, those who first doubted that the sun revolved around the earth, were not only deemed unsound in reason, but "diabolical and wicked," and were visited by the vengeance of the offended good sense of man. Extensive, too, was the sway of the theory of Des Cartes, which preceded the theory of gravitation. The Cartesian hypothesis, representing the worlds as moved by vortices of revolving matter, and the fall of bodies as motion downward by the decrease of the range of the vortex, they sinking to the lowest part of the whirlpool, held for a while an even contest with the hypothesis of gravitation, the battle hanging long with doubtful issue. Because it was decided at last in favor of the Newtonian school, is it to be supposed that the speculations of that truly great man, Des Cartes, were all absurd, that he affirmed no truth, that Newton was right in all his affirmations, and that the power of progress expired with his life? How great were Galileo, Copernicus, Leibnitz, Kepler! Yet who takes for granted all that they asserted? To see truth fitfully, through the obscurity of error, is the most that can be hoped for by any man, or can be claimed as the result of the labors of any philosopher.

Among the minds distinguished as far surpassing in power the range of ordinary intellects, was that of Kepler. He seemed, on some points at least, to have an intuitive perception of truth, and his discoveries have led to extended practical results. His determinations of the laws of planetary motion are, and must ever continue to be, the basis of knowledge in astronomy. Yet even he believed that the earth was an animal, and many of his other ideas create a smile, they appear so fanciful and absurd; but fanciful and absurd because time and further observation have refuted them.

It is no evidence of a want of reverence for the great, that we believe and assert that no individual is without error or has attained the whole truth; that no authority, however great, should determine and settle every principle in philosophy. "There were giants in those days;" but so long as the race of man exists there will be strength.

"Why," said a very great man, far back in the ages of time, "why is it that neither very small nor very large bodies go far when we throw them, but in order that this may happen the thing thrown must have a certain proportion to the agent which throws it? Is it that the thing which is thrown must react against the thing which pushes it; and that a body so large as not to yield at all, or so small as to yield entirely and not to react, produces no throw or push?" Should the human mind have rested upon philosophy like this? By no means. The theories of philosophy are not truth, but monuments set up to mark the progress of man in his search for truth,—guides in the path which leads to the truth. We should read with reverence for the great minds of olden times, while rejoicing in the advance which present opinions indicate; we should read present opinions in the belief that they also will serve future generations to mark a continued and accelerated progress; we should read with respect for those who have faithfully discharged their trusts, and with confidence that those who are yet to do work upon the earth, will do it even more successfully, for having the vantage ground of former attainments,—the highest reach of the minds of the greatest men of one century furnishing the elements of thought to the succeeding age.

Those who reverence Newton should show their reverence by imitating him in his unwearied and eager search for *new* truth. He was not satisfied with philosophy as he found it; but, taking the floating theories of his time, he employed his immense intellect in giving them system, character, definiteness, so that his philosophy, partaking of the vigor of his mind, has so long held dominion over the world. Though his name is most intimately associated in the general mind with the theory of gravitation, his labors in this department used but a small portion of his energy. Every subject in the vast range of subjects to which he directed his attention retains the marks of his strength. He is often thought of as a mere mathematician, verifying the ideas of others. Far from it. His originality of mind equalled his mathematical exactness of conception. How eagerly did he seize upon every new truth, how ready was his mind for progress! The contemplation of such a character should prevent us from ever remaining satisfied with ideas to which the least doubt can be attached.

CHAPTER V.

"IN THE CONSTANT OSCILLATION OF THE HUMAN MIND BETWEEN IDEAS AND FACTS, AFTER HAVING FOR A MOMENT TOUCHED THE LATTER, IT SEEMS TO SWING BACK MORE IMPETUOUSLY TO THE FORMER." — *Whewell.*

WE pass from the examination of the very minute and of the far-extended, to the common facts of our daily observation; gladly throwing down the telescope and microscope, we will look with the naked eye at nature in the ordinary range of vision.

Were we to select one class of facts to illustrate the ideas we present, and to disprove the attraction of matter, it would be that which relates to oscillatory motion, — the action of the pendulum and kindred movements.

The swing of a pendulum, increasing in velocity in proportion to the decrease of distance from the centre of oscillation, the force of its motion being measurable by the area of the circle of which it describes a segment, shows that the nature of force is the same, however its energies may operate. It brings to the mind the relative speed of the planets, increasing as the diameter of their orbits diminishes. Its regular beats mark time with the same precision as does the harmonic motion of the heavenly bodies. It is, therefore, a free, unrestrained movement, showing the laws of force, and had it been rightly understood, the laws which regulate the planetary movements might have been distinctly inferred from it. "We feel," says Bailly, in his History of Astronomy, "that nature is very simple in her operations; the positions and motions of the planets offer at first sight the appearance of intricacy, but the principle which impels them has a naturalness and simplicity like the character of truth;" and this principle has application not only to them, but to all natural and unrestrained motion. Thus in the pendulum, its orbit prescribed and a uniform force of impulse given, motion ensues, invariable, equal, measured by the area described.

A pendulum beating seconds does not change as to time of vibration with the addition or subtraction of weight, nor with a stronger or weaker impulse given to it; the weight of the mass moved and the length of the sweep do not change the time, if the same length of rod is retained. The force of motion is ever in proportion to the mass, and the stronger impulse only increases the range of motion with an increase of velocity that gives the equal time. The impulse lifts it higher and it falls with greater rapidity, and the fall both in distance and velocity determines the succeeding rise. It is evident, therefore, that this result is produced by the same principle which occasions the fall of bodies to the earth; for the downward force is just the force of a body falling vertically the same distance, and the rise is also the same both in relation to distance and velocity. It is the same principle which causes the rise and fall of a stone thrown into the air, to be in equal times. How can this equal rise and fall of the pendulum be explained on the theory of gravitation? Where is the reference to the centre of attraction,—where the *gravitating* power? How is it that the power drawing to the earth draws and repels equally? The rise is unquestionably a continuation of the same motion as the fall, proving that the force has no more reference to a downward than to an upward attraction.

Galileo first called attention to the fact that in the case of a stone swung by a string, the time of its rise was equal to the time of its fall without reference to the path or orbit of its rise; thus pointing to the truth that oscillation is but the transfer and retransfer of the force of rotation, and that, consequently, the force disengaged by the fall is the force which occasions the rise. The force is produced by the descent, and is absorbed by the ascent; it is in proportion to the degree of descent, and is necessary for the same degree of ascent; the degree of force is proportional to the change of level, one mean level being ever preserved in all oscillation. Therefore, the cause of oscillation is to be referred to the moving body and not to the earth.

We repeat the idea; the force by which the oscillating body moves is the force disengaged from rotation by the decrease of the level of rotation. It is the unseen force by which all things move in their diurnal round, for a moment becoming visible in the curve of the pendulum, as the unseen motes floating in the air show themselves for an instant in passing through the beams of the sun.

From the view we have taken it will be perceived, that as the pendulum preserves one mean level, and as its force of descent is, as it were, annihilated by its ascent, it can neither communicate force, nor be the source of any motive power. We can also understand the manner in which the force of impulse acts; it gives the elevation by the descent from which force is produced. Practically, to continue the oscillation of the pendulum, force of impulse, as by the weight of the clock, must be added continually; for a single impulse of elevation is soon lost by friction, and by the force taken to move away the air from the path of the vibrating mass. In a vacuum and without friction, the pendulum would continue to act without a renewal of the impulse. To understand oscillation, we should separate in our minds the *cause* of oscillation, — the transfer of rotative force to other motion and its reapplication to rotation, — from the impulse which first elevates the mass to the position from which it falls, and which is only the *condition* of oscillation.

This view of the subject takes away many of the abstractions which have been connected with the action of the lever, and with the motions of bodies in equilibrio. It throws over all connected subjects a plain and intelligible law. The mind escapes from embarrassment. What more simple idea can we have in relation to the action of a lever of unequal arms, where a smaller weight balances a larger weight and oscillates with it, than to know that one in proportion to the mass ascends in the degree in which the other descends, the greater descent of the smaller weight affording sufficient force for the less ascent of the larger weight, and the converse? How difficult to conceive that the earth, with an attractive force in propotion to its mass, could draw down five pounds so as to elevate ten pounds, merely because the two unequal weights are at different distances from a certain fixed point, though at the same distance from her centre!

In a scale-beam balanced by two equal weights, how could oscillation take place under the law of gravitation? Yet if you give one of these weights an impulse, they will oscillate, — oscillate forever in a vacuum without friction. Gravitation has no power to give alternate motion, — attracting both weights with equal force, it cannot first make one heavier and then the other. If both were held with equal strength, it would be absolutely impossible for this vibration to take place ; there is an absence of all cause, or tendency, or capacity for oscillation. "Vis inertiæ of motion" gives no aid ; for the motion is suspended and renewed at every vibration. Nor does "action and reaction;" for the difficulty is to account for the action. The earth might as reasonably be supposed to attract only one side of an evenly balanced wheel, and thus give it continuous rotation, as first to attract one weight and then another, when both weights are equally heavy. The motion is unquestionably from the transfer and retransfer of rotative force. The balance rotates as one mass with one degree of force, but this force flows from one part to the other of the mass.

So far from being able to induce oscillation, the law of gravitation would immediately overcome the motion. The power of attraction, it is said, *in*creases with the decrease, and *de*creases with the increase of the distance from the centre of attraction. The weight going down is therefore more forcibly attracted, the weight going up is more feebly attracted, and this in an increasing ratio, both for the depression and for the elevation. The difference, it may be said, is so slight, that its results can never be detected by observation. But slight as may be this want of equilibrium, it actually exists. Balances have been made so perfect and so nicely adjusted, as to turn by the impulse of the thousandth part of a grain. Suppose a perfect balance without friction at the fulcrum, and acting in a vacuum. Here this want of equilibrium would be felt; the descending weight, being more attracted, could not rise ; the ascending weight, less attracted, could not fall. There could be no oscillation under the law of gravitation.

Thus nature in the vast range of her adjustments would feel

and respond to this difference of attractive power at the differing distances from the centre of attraction. Throughout extended space there could be no oscillatory movement. The spheres, according to theory, evenly balanced in their orbits by the central and tangential forces, would, by mutual attraction, have erratic movements. There would be perturbations, actual perturbations, — wanderings, actual wanderings, from their path under the increasing power of the lessening distance. Their changes could not be compensatory and periodic; the mean time of revolution and mean extent of axis of orbit could not be preserved; they would aberrate, but never oscillate, for the force causing aberration would be increased by the act of aberration.

On this earth too, under the law of gravitation, the order of nature would be broken up. Off the Cape of Good Hope, there are at times waves more than a half mile in breadth, and very many miles in length. Millions and millions of tons of water are thus heaped up in one volume; the increase and diminution of attraction in this case would be felt and responded to, for there are out of balance thousands of tons. It must be that the ascending water would continue to rise, and the valley of depression to sink lower. What too would be the result of the great tidal wave raising and depressing the waters of an ocean? Quietly to sink the less attracted elevation, and abruptly to lift the more strongly attracted depression? Impossible; under the law of gravitation there can be no bound or limit, no oscillation whatever. Perturbation, swerving motion, is without check. There is no voice of command, Thus far and no farther shalt thou rise, and here shall thy proud waves be stayed.

But on the other hand, in the hypothesis we offer, oscillation is measured in its extent by the impulse. The force of the undulation is neither an increasing nor a diminishing force. It is not acquired from other bodies, nor imparted. It exists in the mass moved, and is transferred from part to part. It is a current of force flowing by an impulse. This idea will be further elucidated as we proceed, and a principle will be unfolded by which the oscillation of all fluids is limited within a definite range, so that the equipoised and freely moving element, water, has a fixed bound to its oscillations.

We would rest the truth of the hypothesis on this nice adjustment, this beautiful equipoise of motion by which is secured the permanence and harmony of creation; and with these ideas, the rise of the wave reflects from its spray not only the light of the sun, but its motion reflects to the mind a more beautiful light,—light revealing the law of nature, by which movements as free as the breath of the breeze are held in subjection with a chain so strong, that He only who forged it can rend it asunder.

We will now recur more particularly to the phenomena of the waves. To show that they are not distinctly understood, we will cite the following description from a standard work on Natural Philosophy:—"Certain particles of water are first forced down, and the surrounding particles are forced up above their level; their circular ridge subsides, and not only fills up the original depression, but from its momentum forces up another ridge exterior to it; this subsiding forces up another ridge, and so on." This description calls to mind the following passage from an ancient philosopher:—"His whole performance seemed to reach no further than if a man should say, 'Socrates does all by intellect,' and after proposing to give a reason for my action should say, 'I am sitting on my bed, because my body is composed of bones and nerves; the bones are hard, solid, and separated by joints, and the nerves being able to bend and unbend themselves tie the bones to the flesh, &c., and *that* is the reason forsooth that I sit in this position.'"

But we need an explanation of the phenomena of the waves more distinct than comes from the mere recital of the phases of their rise and fall. It is expressly mentioned by Plutarch, and by Pliny, that the seamen of their day used to still the waves by pouring oil into the sea. A letter written in 1707, speaks of a storm at sea "which had nothing particular about it, except that the captain found himself obliged to pour oil in the sea, which had an excellent effect, and succeeded in preserving us." Franklin, in his paper on this subject, says: — "A small quantity of oil, for instance, a quarter of an ounce, will spread itself quickly on the water in a pond to the extent of an acre, and if poured on the windward side, the water will thereby be rendered quite smooth as far as the film of oil extends. . . . . . Its principal operation is to prevent the rising of new waves, and to prevent the wind from driving those already raised with as much force as it would if their surface was not oiled."

This idea of stilling the waves by means of oil is omitted in recent works on the subject, and is considered by many as a popular superstition; but when was Franklin mistaken in facts which he investigated? and what popular belief is there thus extensive which has no foundation in truth? Why is it, as before stated, that a thin stratum of oil spread over the surface of water, will prevent its freezing when exposed to a degree of cold much below the point of congelation? Why does oil prevent the adhesion of substances pressed together? All these results are from the non-conducting properties of oil. Force as heat does not escape through it; force of cohesion does not pass this barrier; force to give the impulse which occasions the undulation of the waves, is repelled from the water. The theory of the identity of force *a priori*, without experiment, would have indicated these results.

A fact analogous to the rise of the wave, is the rise of the air in a ventilating tube; over the upper orifice passes the horizontal wind, particles of which, by their friction or attrition on the air in the tube, are arrested in their course; the force by which they were impelled is imparted to the air in the ventilating tube, and consequently the air is raised to a higher level of rotation. Wherever there is a passage of a current of matter, any part of which is arrested in its course, that which arrests it is moved. With this idea fixed in the mind we are able, not only to give a satisfactory explanation of the rise of the wave, but also of many common facts coming under our daily observation.

When the wind passes over the surface of the sea, particles of this ærial current are arrested in their course by any projecting inequality, or in the absence of already formed waves, by friction or attrition, as already explained. The water has additional force transferred to it by the passing wind. It acquires force for rotation at a higher level. It therefore rises. The elevation of the first wave, however slight, offers a greater impinging surface; it therefore continues to rise, and its rise is proportional to the strength of the wind which causes it. This undulation of the water arises from the same principle as the oscillation of the pendulum, the motion of which is the reception and transfer of the force of rotation at its different levels; the descent liberates force for the ascent, and one mean level of rotation is ever preserved. Thus rises the wave, its motion the reception and transfer of force, its impulse of oscillation the force robbed from the passing wind. Thus, and thus only, can we explain a fact which is rather indefinitely expressed in the following language from high authority: "Momentum may be accumulated to an enormous degree in very large suspended masses, by forces, which, at their first application, appear to be totally inadequate to move them." The impulse *is* "totally inadequate" to cause the motion; a child's hand cannot move a thousand pounds, but the child's hand can destroy a balance, and the force of nature transferred according to the level of rotation is the cause of motion. In the case of the wave, the wind destroys the equilibrium; but the action of a more intense force moves the mountain of waters.

The motion of the wave is deeply interesting in all its aspects. The crests of the surge rise vertically, though they are apparently progressive. It is not until the water shoals toward the beach that they acquire a progressive motion, — progressive because near the shore there is not depth for the vertical movement. It requires double the generally supposed depth to form the vertical wave; for the uplifted water falls as much below the valley of depression as it rises above it. There is a descending wave, a current under water, which as a wedge forces up the succeeding wave, an opposite wave under the water. Hence on the shoaling water is the undertow, a retreat of the water at a certain depth with advancing water at the surface. The extent of this reciprocal wave is in exact proportion to the extent of the visible wave.

Besides, the surface of the atmosphere is also to a degree in oscillation with the water over which it is spread. The wind conforms to the surface, it oscillates with the water; the sails of a boat, or the lower sails of a ship, are not becalmed with a steady wind, though surrounded by a wall of water. Says an experienced navigator, "the lower sails in scudding are at times becalmed, from the rise of the stern of the ship, but never becalmed *on* the wind, though the waves rise higher than the sails, a fact which I have often tried to account for." For this reason, — the undulating motion of the wind over the undulating surface of the sea, — is it that the wind off shore, not having acquired the corresponding undulation, however strong it may blow, smooths the sea.

A most interesting scene was described by a friend, who stood on a small rocky island in the midst of the ocean, looking with intense delight on the forces of nature as displayed in a storm. The mountain waves were dark, almost black, the intensity of their gloom being heightened by the circle of white foam which surged upon the rocks. A sudden veering of the wind, this wind being without the conforming undulation, prostrated the billows as it were at once; the ocean became as smooth as in a summer's breeze. There was no surge on the rock; but the surface of the sea was white with foam and curling in wreaths of vapor as far as the eye could reach. Gradually, however, the new wind acquired the sympathetic undulating movement, the waves again began to rise, and soon the surf dashed with its former fury against the island.

Only because the rise and fall of the wave are from the reception and transfer of force, the mean level of the ocean ever remaining the same, could this result have been produced. If there had been "accumulation of momentum to an enormous degree," — if the attracting earth had drawn down the depressed waters with more strength than the elevated waters, — the ocean could never have been thus smoothed into the quiet of one level; the new wind would have increased the surging waves, so that they would have lifted their crests still higher in confused and broken masses, resulting in a conflict of motion which the strength of no vessel that ever floated could withstand. There is something fascinating in the contemplation of the intensity of the force of the elements so perfectly under the control of law, as is witnessed on the barren rocks in the midst of the ocean's storm.

Before passing to other instances of oscillatory movement, we will notice some analogous phenomena. The whirlpool and the whirlwind are illustrative of the action of force. We will copy first a description of these phenomena which contains their usual explanation. "When gusts of wind come from different quarters, and meet in a certain place, there the air acquires a circular, or rotatory, or screw-like motion, either ascending or descending, as it were, around an axis; and this axis is sometimes stationary, and at other times moves in a particular direction. This phenomenon, which is called a whirlwind, gives a whirling motion to dust, sand, water, a part of a cloud, and sometimes to bodies of great weight and bulk, carrying them upwards and downwards, and at last scattering them in every direction. The water-spout is formed by the centrifugal force of the wind, creating a vacuum in the centre through which the water ascends." These are the common ideas; for circular motion is usually considered as the result of conflicting forces.

The action of two currents of water, in the formation of the whirlpool, and of two winds in the formation of the whirlwind, is precisely the same. The onward motion of the particles of water or air is arrested by the impingement of the two currents, but the force which impelled them is not annihilated; the onward motion only is ended, and the force from its nature must produce motion. The circular is the only open range of action; for no separate particle of either stream can be accelerated in the stream by a sudden impulse. Therefore, the arrested water or air moves in a vortex until its force is gradually imparted to the current in which it floats.

In the water-spout a part of the force of the revolving wind is conveyed to the water; the water, having force for rotation at the higher level thus communicated to it, rises. The water on which the whirlwind rests becomes violently agitated with the additional force received, and is elevated as foam, vapor, and water and air mingled together. Of one an observer said, "the water at the base seemed to boil and go up in detached masses mingled with the air." The raising of heavy bodies on the land and of the water of the sea by whirlwinds, in fact, the elevation of matter generally, is ever on the condition that the due degree of rotative force is supplied. It may be supplied from below, as in the act of pushing up; it may be supplied from above in the act of drawing up. All the phenomena of change of level are explained on this principle; the falling body imparts its no longer needed force, the rising body demands an increased supply.

This idea is finely illustrated in the fall of water from a great height into a basin of stone, as in the "Pool" at the White Mountains. The water received in the basin descends from a great elevation; it has great spare force of rotation, which force, not transferred to the rock on which it falls, impels the water in a vertical movement, and thus forms for itself a circular receptacle. If the water had barely flowed into this receptacle without force, or from a moderate elevation, the issuing water would have used this force, a current between the inlet and outlet would have been established, and no vertical movement been induced. Water arrested between two currents makes the whirlpool. Arrested water from a great fall only, can make a permanent whirlpool, which will scoop out for itself the *vortex mould* from the solid rock.

A fact which has strayed somewhat from all theories, and which the books on Hydrodynamics that we have examined do not even notice, — a trivial fact, coming under our daily observation, — is fully illustrative of the principle; we refer to the vortex in the body of a tunnel full of water, which is formed on the issue of the water from the pipe below. Fill a tunnel with water; when it is stationary allow it to flow out from the pipe; the water in the tunnel immediately assumes a vertical movement, and whirls faster and faster as it descends.

We have inquired of many who were well read in the usual explanations of philosophy, and their answer has been that the cause of this vortex is the conflict of two forces. How is this? By theory, there is only one force present, the force of gravitation, the downward drawing force of the attraction of the earth. One suggested that the vortex was formed by reason of the conical shape of the tunnel; that he was wrong is evident from the fact, that a tunnel of any form gives the same result.

The explanation of the fact is very simple on the principle which we present. The water issuing from the tube uses, in its downward motion, all the spare force of its own descent to a lower level. But the water remaining in the tunnel has also descended to a lower level of rotation. It has therefore spare force, and this force has no other range than to confer circular motion to the water. The water therefore revolves. Thus is it, that facts, common, trivial facts, declare the general law.

We return from this digression to oscillatory movements, and affirm that the surface or crust of the globe, as it has been called, has undulatory motion.

We think that we shall be able, as we proceed, to establish the fact, that the barometer indicates, not the density of the atmosphere and consequently the level of rotation, but that it indicates this level positively and directly. For the present, however, take the barometer as it is commonly understood, as indicating by the oscillations of the mercury the density of the atmosphere.

There are oscillations of the mercury entirely unaccounted for by this theory. At the same apparent level the barometer has a range of about three inches. This well known fact cannot be explained by any supposed variation in the density of the incumbent atmosphere. Philosophy acknowledges her ignorance of the cause of these changes. The dryness or the humidity of the air will not account for them; this is admitted. Winds cannot give the reason; for it would require the passage of a current of air two thousand eight hundred and eighty miles, at the rate of sixty miles an hour, to account for a variation of one half inch of the mercury. Besides, these changes often precede the change in the atmosphere; they may take place when there is a dead calm, or during the blowing of a steady gale, thus foretelling atmospheric changes rather than being produced by them. Do not these facts indicate that the same cause which changes the level of the barometer produces the change in the atmosphere, not that the atmospheric change affects the barometer?

These oscillations of the mercury are not without cause, and there is one cause to which we would assign them, — the rise and depression of the earth's surface on which the barometer stands. We believe that the earth is to a degree elastic, yielding; that it is susceptible of undulatory movement, — periodic, or occasional, — so extensive in range, yet so slight in amount as to produce no convulsion, no apparent change; yet to be traced by the oscillations of the barometer, and by the winds and storms which indicate a new distribution of the force of rotation.

A scientific writer says, that the surface of the globe is no more stable or firm comparatively than is a heap of saw-dust which is floating on the surface of the water. This is a strong metaphor, but he drew his inference from a supposition which we are far from believing, — from hidden fires shut up in the bosom of the earth. For, as he says, if in the same proportion the heat continues into the interior of the earth, which is evidenced by the fact of its increase so far as we have penetrated, at one hundred miles from the surface it would be greater than any artificial heat ever produced. The most refractory substances would be reduced to a fluid state; therefore, the solid crust is of small comparative depth, and the internal fires by their intensity may well be supposed to give motion to it.

But there are many well known facts which prove that there are fluctuations of the surface of the globe. An elevation of the coast of Chili of eighty-five feet has been proved by geologists, and a recent elevation of four feet, extending along a thousand miles of the South American coast, was produced in a single night. Earthquakes have been felt throughout an entire hemisphere. During the great earthquake near Lisbon, the waters rose with the upheaving earth one hundred feet, and the motion was propagated across the Atlantic Ocean, vessels feeling the surge according to their distance from the centre of motion. We would also refer to the "periodic although irregularly alternating rise and fall of the waters of the Caspian sea," and to the fact that harbors in the Mediterranean sea have been suddenly left dry for many hours. It is stated by an eminent geologist, that the strata of the earth, her ridges, her mountains and valleys take the general form that would be induced by continuous undulations; and geologists generally assert the same fact, referring it to various causes. If there be changes which result in convulsion, and which thus leave a permanent record, why may there not be gentler changes, fluctuations unattended with severe catastrophes or abiding results? We have the authority of Humboldt for the assertion, that without earthquakes the surface of the earth is capable of gentle and progressive oscillations; and he also remarks that "*the opinion so implicitly entertained regarding the force of gravity at any given point of the earth's surface, has in some degree been controverted by the gradual rise of large portions of the earth's surface.*"

The sudden changes of level in the waters of our great lakes have long excited attention, and have never been satisfactorily accounted for. We copy from the Annual of Scientific Discovery a condensed account of the facts so far as they are known, and the hypotheses which are offered for their solution. "The results of recent observations prove that these waters do not rise and fall at stated periods, corresponding to the ebb and flow of the tide, but are subject to extraordinary risings, which are independent of the influence of the sun and moon. These risings attracted the attention of the earliest *voyageurs* in these regions. Charlevoix, who traversed the lakes nearly a century ago, says in reference to Lake Ontario:—'I observed that in this lake there is a sort of reflux and flux, almost instantaneous; the rocks near the banks being covered with water and uncovered again several times in the space of a quarter of an hour, even if the surface of the lake was very calm, with scarce a breath of air. After reflecting for some time on this appearance, I imagined it was owing to springs at the bottom of the lake, and to the shock of their currents with those of the rivers which fall into them from all sides, and thus produce those intermitting motions.' The same movements were noticed by Mackenzie, in 1787; by an expedition under Colonel Bradstreet, in 1764; on Lake Erie, in 1823; and at various later periods. In the summer of 1834, an extraordinary retrocession of the waters of Lake Superior took place at the outlet of Sault Sainte Marie. The river at this place is nearly a mile wide, and in the distance of a mile falls 18.5 feet. The phenomena occurred about noon. The day was calm, but cloudy. The water retired suddenly, leaving the bed of the river bare, except for a distance of thirty rods, and it remained so nearly an hour. Persons went out and caught fish in pools formed in the depressions of the rocks. The return of the waters is represented as having been very grand. They came down like an immense surge. In the summer of 1847, on one occasion the waters rose and fell at intervals of about fifteen minutes, during an entire afternoon. The variation was from ten to twenty inches, the day being calm and clear; but the barometer was falling. Before the expiration of forty-eight hours a violent gale set in. At Copper Harbor, the ebb and flow of the water through narrow inlets and estuaries have been repeatedly noticed when there was not a breath of wind on the lake. Similar phenomena occur on several of the Swiss lakes. Professor Mather, who observed the barometer at Copper Harbor during one of these fluctuations, remarks:—'As a general thing, fluctuations in the barometer accompanied fluctuations in the level of the water; but sometimes the water-level varied rapidly in the harbor, while no such variations occurred in the barometer at the place of observation.'

"As a general rule, these variations in the water-level indicate the approach of a storm, or a disturbed state of the atmosphere. We are therefore led to infer that these phenomena result, not from the prevalence of the winds acting on the water, accumulating it at one point and depressing it at others, but from sudden and local changes in the pressure of the atmosphere giving rise to a series of barometric waves."

How could there be these "sudden and local changes" in the pressure of the atmosphere, with a calm day, and of course an equable temperature? Or how could they exist at all in a fluid so elastic as the air, the very nature of elasticity being such as to correct at once any difference of density in the same volume? Our views lead us to regard these changes in the level of the water as indicative of changes of the level of those portions of the earth on which the water rests. And as a proof that vibration of the earth's surface does occasion barometrical change, we need only refer to the well-known fact that at the time of the great Lisbon earthquake, the mercury fell so far in the barometer even in Great Britain, "as to disappear from that portion of the top usually left uncovered for examination." Humboldt also remarks that "the connection between the eruption of a small volcano with the state of the barometer is generally recognized, although our present knowledge of volcanic phenomena, and *the slight changes of atmospheric pressure accompanying our winds*, do not enable us to offer any satisfactory explanation of the fact."

But we need not accumulate facts. Science generally indicates a change of level of different portions of the crust of this globe, taking place either periodically, or as it were incidentally. Nor are we limited to science for our proofs of mutation. Science may point to the barometer as indicative of these changes; she may show the earthquake and the storm; she may declare, from the upraised strata, from the mountain ridges, from the depressed valleys, from the confused commingling of the primitive, the transition, and the recent formation, that the earth is not an unyielding mass, bound together by a law which gives increased density according to the approach to its centre.

The world, now permanent as the home of man, has passed through at least one great and sudden change. The upturned strata, the drift from the poles so enormous in quantity, that were it again collected and restored it would change the form of the earth, abundantly prove this. We know that this great change has been accounted for by natural causes, by the accumulation of the present changes in foregone times, — by the condensation, as it were, of present events in the lapse of past ages, — by the volcano, the earthquake, the storm, the drift of rivers, the gradual deposit of floods; — glaciers have been launched, and icebergs freighted with diluvium and boulders. We are not prepared to seek a solution of the great changes of the earth from natural causes, from the intensification of the present action of force, condensed to stronger action in the youth of the world. There has been one change at least, which came not from a natural cause, from the every-day phenomena of nature. We refer to the time when the earth was without form and void, and darkness was on the face of the deep, when the spirit of God moved on the face of the water, and it was said, Let there be light, and the rays of the sun created "in the beginning" for the first time penetrated the mist that went up from the surface of the earth; when "the firmaments rose" and "divided the waters from the waters."

How great the change which thus converted a chaotic world into a bright and beautiful home for man! The intensity of the force is written down on the broken strata, the torn and ragged mountain ridges, and in the smoothing and furrowing of the stubborn rocks. The form of the earth was changed, and with this change came necessarily a change of its rotation, and it may be of the form of its orbit; for a change in one condition of a sphere, must change its every phase.

And that this change did take place proves that there can be no reference to any supposed centre as the radiating point of its motive force; for the globe is self-poised, asking of no other leave to be, borrowing of no other the power to move,—connected, as a part of a system of worlds, yet speeding on in a path of her own in perfect balance, in accurate equilibrium, as if buoyed up by the right arm of God.

As there is a perfect equipoise in the earth as a sphere, as a whole, so is there the same perfect equipoise in her parts, in her elements, in her constituting masses. From this perfect equipoise come the slight aberrations and perturbations, the oscillations of the surface. As the pendulum swings on the condition of equilibrium ever preserving one mean level, so may the surface vibrate of the even-balanced earth. Therefore is there freedom, ease, elasticity of motion, — nothing stationary, yet nothing wandering to an extent that can mar the designs of creative wisdom; all things flowing gently as if without rule, yet bound together by laws as sure and as unchangeable as is the power of Him who formed the earth in the beginning, as the home of his children, whose security and happiness were the objects of creative skill.

An idea given in a sermon by the Rev. A. P. Peabody has fastened itself in our thoughts; we quote from memory: "In the Ptolemaic system, this earth was considered as the centre of the system to which all the wandering hosts of heaven were tributary, — aids, accompanying circumstances. The better lights of astronomy dispelled this error, and the earth is looked upon now as a part, a minute part of a universe without a conceivable limit. But the earth from this is not degraded, but exalted, — not decreased, but increased in interest. The part of such a whole far transcends in sublimity the centre of a system so narrow that all refers to its own limited bounds." The idea extended to man imparts to the individual a dignity proportional to the extent of the universe of which he is a participant, — little in himself, yet how great considered as an intregral part of the whole creation! And this intimate connection, too, not merely appertains to his physical nature, but he has yet greater dignity of position from the view that, connected as he is with God's material universe, there is also for him a connection with the Sustaining Spirit, the Life of the Universe, — that indeed he is the child of God.

We return from this "oscillation" of thought. In a

still night, at a hotel of great extent, built of massive granite, a vibratory motion was distinctly perceptible. This great stone building, the adjacent buildings of brick, and the intervening ground to a great extent were perceptibly moved by the passing of a carriage over the paved street. Did the muscular force of the horse thus produce motion of the enormous mass of stone, brick, earth, and the solid ledge? The impulse only was given, and this impulse, slight as it was, so nicely adjusted is the balance of force, resulted in oscillation. The vibration called forth by the motion of the carriage, was the bringing into action the weight of the mass,—the transfer and retransfer of the force of rotation as the oscillating matter rose and sunk in level. Had gravitation bound down the enormous mass, drawn it with the force of thousands of thousands tons, it would have stood firm and unyielding. How much safer for a ship to float on the free moving water, than to receive the shock of the storm when stranded on the solid rock!

As we write, we raise our eyes to a range of lofty mountains, the top of one of which rises some four thousand feet, with a summit of bald rock, the granite reaching from this height to the broad base, and extending for many long miles below, ledge upon ledge, stratum upon stratum, all bound together by the circulation of force through every particle, fragment, mass. But there is no weight. The granite crystals touch each other no more closely at the lowest base than at the summit, nor press more heavily on each other. There is no weight, except on the condition that the mountain descends to a level of rotation. It is self-poised, as is the whole earth.

Practical men, who know only what they have seen, and

believe in what takes place, without being troubled by theory, work on the faith that there is no force of gravitation in the mass which preserves one line of rotation. In taking away the foundation of a brick wall which had been undermined, to rebuild it, the owner expressed his fears that too much had been removed of the support at once, and that the building would fall. The reply of the mechanic was, "I used to fear, but now I know that it is safe until it *begins* to move, and I shall work without any jar; if it should begin to move, twice the present support would not keep it up." In taking away the support of the arch of a stone bridge, after the keystone was put in, a slight, sudden sag of the bridge crushed the keystone, — a stone, the cohesion of which, if standing firm at one level, would have borne all the granite that could have been piled miles high upon it.

This principle is evidenced by the superior security of the arch for bridges and similar structures. In the arch, there can be no descent without crushing the material of which it is formed. Thus too has the passing of a body of troops over a bridge, with measured step by beat of drum, often given the vibration which calls into action the weight or force of descent. In the Britannia tubular iron bridge of immense weight, some of the spans of which are four hundred and seventy-two feet in length, care has been taken to secure the bridge against vibration; the upper portions of the tube are made thicker, operating as an arch, the particles of which are compressed by any descent. Heavy as it is, it will undoubtedly stand if it has no vibration to call forth the force of descent.

The moon floating around the earth will not fall because she possesses force for rotation, in proportion to her distance from the earth. So the crust of the earth cannot fall, as it ever has the due force for rotation; there may be oscillation in the orbit of the moon, the mean distance being preserved; there may be a movement of the earth's surface, but it is an oscillatory movement; there is both the rise and the fall, one mean of rotation being preserved. Masses and fragments of the earth fall, when unsupported they take a lower level of rotation, and as they fall yield up to other bodies the force of their former superior level. Hence every movement is oscillatory, the transfer of force, — the force remaining of fixed quantity in all its transfers, preserving the perfect equilibrium of creation.

After the assertion of the law of falling bodies by Galileo, there was found a great difficulty in applying it to the commencement of the motion; for this beginning of motion clearly shows that they do not fall by attraction. Great disputes took place in regard to the velocity with which bodies begin to fall. It led to the consideration of the infinite divisibility of space; for the motion at its commencement is so slight that the body hardly appears to move, or, as it has been expressed, "when a body begins to fall from rest, it begins to fall with *no* velocity. In one thousandth part of a second it has only acquired one thousandth part of the velocity which it has at the end of one second." If gravitation act with its supposed force upon a body at rest, how great would be its instantaneous velocity when the support was taken away! Its fall would be as sudden as the recoil of a bent spring.

Nor could it be understood why bodies which had fallen by the force of gravity should be able to rise again to the same level against the force of gravity; or why on the inclined plane the velocity of a body should be measured by the vertical height from which it falls; nor why the fall should be with uniformly accelerated velocity. Even Galileo, vigorous as were his powers of thought, could not master the subject. And the obscurity has continued to this day.

The theory of gravitation acknowledges a difference between the weight of a body, that is, its tendency to fall, and the force of the fall, — weight being an invariable quality, while the force of descent is in proportion to the time of the fall. We regard the cause of the fall, and the force given out by the fall as essentially different, and we hope that we shall be able, in the process of this inquiry, to explain the cause as well as the results. But whatever it may be, the stone unsupported will not remain at rest. We have just read in an account of a recent ascent of Mont Blanc, that "spots have to be passed where no word can be spoken lest thousands of tons of snow should be set in motion;" for it has been known, that the mere vibration of the air, caused by the human voice, has brought down the avalanche. The intensity of molecular action, the pulse of nature, the very throb of force, the vibration and oscillation of all things would account for the fact that the suspended body has motion, has the beginning of its downward course. The impulse given, the most minute line of descent induced, there is the spare force of rotation, there is force for the motion of the mass in a new direction. It has commenced its fall, — it has unused force. It cannot rotate more rapidly at the lower level; for it is a part of a current of matter passing round the earth in an equable stream. Downward, then, it must take its way. Its very appearance in the act of falling indicates the law of its motion. It is not attracted; for it does not rush at once to the earth, as do the particles of steel to the magnet; but, for an instant, as if in suspense, like

> " A long-swept wave about to break,
> It on the curl hangs pausing;"

then, gathering force for its downward career, it falls;

> " Still gathering force it smokes, and urged amain,
> Whirls, leaps, and thunders down, impetuous to the plain."

The force obtained by the change of the level of the falling mass,—of water for instance,—cannot come from the attracting earth; for the attractive force is force to draw the mass toward her centre, and having acted according to this, the essential principle of attraction, it would cease to act, or merely hold the water at its new level. But the force of the fall of bodies remains; it is a living principle. It was in them at their higher level, there acting in rotation. If water falls, it transfers this force to other bodies, inducing motion in them. If not imparted, it is sufficient in intensity to raise the water again to its former level. It is not a mere gravitating, or downward force, but force giving motion irrespective of direction.

Thus, the fall of water over an artificial dam is not from the downward attraction of the earth. That force can only exert its essential property of bringing the particles of matter nearer to each other. In doing this, it has discharged its only conceivable function. But the particles of water by the act of descent impart the spare force of their lower level of rotation; and this force, diverted from the majestic stream of rotative power, can have given to it by mechanical appliances the line of direction which crushes the grain, or on the revolving spindle twirls the fibre of cotton,—the mechanic using the force of nature for the comforts and luxuries of life. Thus is imparted to the practical mechanic a dignity arising from his position. He is not a mere cunning workman in wood and brass; but he bears in his hand the sceptre of power over nature, — he controls her energies, — he borrows her force, — from water, wind, and steam, and through the lever and the wheel, he directs the action of this force for the well-being of man.

When engaged in the examination of facts, the thoughts will oscillate from them to abstract principles, — from material things to that which they represent. Does not this law of the mind prove that the connection of the intellect with material objects is not its only tie to the universe? We have considered the artisan as borrowing the power of nature for his purposes, — considered him as not confined to that degree of strength only which flows through his own frame; he goes out of himself to the great fountains of energy for his purposes. So, he who would comprehend the things around him, — he who is the artisan in philosophy, — is he confined to his own feeble strength, limited to his own power of thought? Is there not a communion between him and truth more direct than comes from the contemplation of external objects merely? Is he never allowed to borrow wisdom from the Fountain of Wisdom?

"Now all the knowledge and wisdom that is in creatures, whether angels or men, is nothing else but a participation of that one eternal, immutable, and uncreated wisdom of God, or several signatures of that one archetypal seal, or like so many multiplied reflections of one and the same face, made in several glasses, whereof some are clearer, some obscurer, some standing nearer, some further off."

CHAPTER VI.

"SURELY IT IS A WORK WELL DESERVING OUR PAINS, TO MAKE A STRICT INQUIRY CONCERNING THE FIRST PRINCIPLES OF HUMAN KNOWLEDGE, TO SIFT AND EXAMINE THEM ON ALL SIDES; ESPECIALLY SINCE THERE MAY BE SOME GROUNDS TO SUSPECT, THAT THESE LETS AND DIFFICULTIES, WHICH STAY AND EMBARRASS THE MIND IN ITS SEARCH AFTER TRUTH, DO NOT SPRING FROM ANY DARKNESS AND INTRICACY IN THE OBJECTS, OR NATURAL DEFECT IN THE UNDERSTANDING, SO MUCH AS FROM FALSE PRINCIPLES WHICH HAVE BEEN INSISTED ON, AND MIGHT HAVE BEEN AVOIDED.

"HOW DIFFICULT AND DISCOURAGING SOEVER THIS STEP MAY SEEM, WHEN I CONSIDER HOW MANY GREAT AND EXTRAORDINARY MEN HAVE GONE BEFORE ME IN THE SAME DESIGN, YET I AM NOT WITHOUT SOME HOPES, UPON THE CONSIDERATION THAT THE LARGEST VIEWS ARE NOT ALWAYS THE CLEAREST, AND THAT HE WHO IS SHORT-SIGHTED WILL BE OBLIGED TO DRAW THE OBJECT NEARER, AND MAY PERHAPS, BY A CLOSE AND NARROW SURVEY, DISCERN THAT WHICH HAD ESCAPED FAR BETTER EYES."—*Bishop Berkeley.*

THERE are, unquestionably, two different casts of mind,—one giving greater, the other less attention to facts; the one the more rigid and exact in observation, the other the more prone to speculation. Hence the perpetual conflict between what are called the practical and the theoretical. But the dispute concerning the values of the two methods in philosophy is in words merely. The man who only knows facts, and the man who, ignorant of the doings of nature, dreams rather than judges, are both unfit for philosophy. Theories unsupported by facts are worthless, and an accumulation of facts without reference to the laws of their combination is also worthless.

10 *

But every man is in some degree a theorist, and he is a theorist because he is a man. The animal alone rests contented with the facts presented to the senses. In proportion to the strength of a mind is its tendency to go beneath the surface of things, to be dissatisfied without hypothesis to connect the facts that are observed, and therefore to seek some principle of unity. Thus every philosopher is necessarily a theorist. Even Bacon, the father of the Inductive Method, was a theorist, and at times a wild and fanciful theorist. "Some noises," says he, "help sleep, as the blowing of the wind and the trickling of the water; they move a gentle attention, and whatever moveth attention without too much labor, *stilleth the natural and discursive motion of the spirits.*" "What indeed," asks a profound and accurate thinker, "are Newton's Queries but so many hypotheses which are proposed to philosophers as subjects of examination?" "And," he continues, "did not even the great doctrine of gravitation take its first rise from a fortunate conjecture?"

It is remarked by Sir David Brewster that "the influence of the imagination as an instrument of research has been much overlooked by those who have ventured to give laws to philosophy. This faculty is of the greatest value in physical inquiries; if we use it as a guide and confide in its indications, it will infallibly deceive us; but if we employ it as an auxiliary, it will afford us the most invaluable aid." After an hypothesis has been suggested by the imagination, it should become the subject of rigorous examination, and be retained or discharged, as it may or may not give the desired explanation; for the proof of the truth of a theory is its harmony with facts. True philosophy, therefore, uses both reason and observation in making progress.

And while philosophy demands that no theory shall be admitted as proved any further than it is supported by facts, and that science should be guarded from wayward fancies and day-dreams, equal care should be taken to reëxamine the theories that are current, so that the errors that must necessarily be handed down in the regular transmission of science, from the education of one generation by the preceding, may be expelled. There is far less danger that new errors be received than that old ones be suffered to remain; for against new theories the educated contend most strenuously, if they conflict with a system to the full comprehension of which a lifetime has been devoted. Errors once established therefore become permanent; they are incorporated into a system. Common sense, which might detect them, defers to the authority of science, and science to the wisdom of former times. However, all things on the whole are wisely determined so that progress shall be made, though it be by a succession of waves surging at long intervals, rather than by the steady advance of the tide. But we have been drawn from our path. We return to the examination of the theories relating to the differing densities of the atmosphere, as indicated by the rise and fall of the mercury in the barometer.

It is generally supposed that the atmosphere increases in density from its upper surface to its line of contact with the globe. In the language of Sir John Herschell, "when we have ascended to the height of one thousand feet, we have have left below us about one-thirtieth of the atmosphere; at ten thousand six hundred feet of elevation, rather less than that of Ætna, we have ascended through about one third; at eighteen thousand feet, which is nearly that of Cotopaxi, through one half of the material, or at least of the ponderable body of air incumbent on the earth's surface." This lessening density as we go up from the surface, is supposed to be proved by the depression of the mercury in the barometrical tube.

Before the experiments of Torricelli, the ascent and support of fluids by a vacuum were accounted for by the abhorrence of nature for a vacuum. When it was discovered that this "abhorrence" was limited for water to about thirty-three feet, for mercury to about thirty inches, and for other fluids in proportion to their density, some new theory became necessary, and the weight or pressure of the atmosphere was substituted. Galileo, Torricelli, Pascal, seem jointly to have formed the new theory, and Pascal fully established it by experiments, which gave the fact that at different altitudes the level of the mercury varied, — that it rose higher in the vacuum as the level of the instrument decreased, and that it fell as the barometer was elevated.

Thus the "*fuga vacui*" being abandoned, the support of the fluid being attributed to the pressure of the atmosphere, and the degree of support being found to lessen upon ascent, the hypothesis that the weight or pressure of the atmosphere decreases according to elevation was established, and remains unquestioned to this day. This theory, however, though it has been useful for many practical purposes, does not give a complete explanation of the facts connected with the barometer.

How is the air supposed to be packed according to barometrical indications? We will quote from one who very distinctly gives the commonly received opinions. "The first stratum of air gravitates with its own share of the earth's attraction, the second with its own share, and with the weight of the stratum over it, and so on; the lower stratum bearing in addition to its own weight all the weight of the air over it." The atmosphere has also been compared to a series of fleeces of wool, the lower the most compressed.

Gravitation of the atmosphere, if the air be attracted at all, would appear at first sight as if threads were attached to each particle, drawing every one with equal tension. No particle could rise relatively to another particle, if all were drawn down with equal force. The particle going down must be more strongly drawn, the particle going up must be drawn with less comparative force. With equal attraction there could be no relative change of place. The only difference in weight of the different strata would arise from the increase or diminution of the force of attraction, according to the distance of each stratum from the centre of attraction. If this force acted on the whole atmosphere as one volume, the whole volume would be of as uniform density as a mass of granite. If the force of attraction operated on each particle according to its distance from the centre of attraction, the difference in the weight of different strata would be far less, than that apparently indicated by the barometer.

Such a condensation of the atmosphere as is supposed could not result from the law of gravitation; for that law assumes a decrease of force proportional to the square of the distance. A volume of one hundred cubic inches of air weighs but thirty-one grains,—is attracted by a force the value of which is only thirty-one grains. How large must be the volume of air which weighs a ton,—is attracted by a force measured by one ton? It would cover the whole of a large and high mountain. This ton of air would weigh but one ounce less, a mile from the surface of the globe, than it would at its very surface. This is the only ratio in which gravitation could change the weight of the different strata of air. The earth can have no favoritism. She holds in place or draws down every particle of air equally, — or, more accurately, in proportion to its distance from the centre.

But, admitting that each stratum of air bears not only its own share of the earth's attraction, but also the weight of all the strata over it, still this could not condense the atmosphere to the degree that is supposed.

No one doubts that the temperature of the atmosphere changes with its level. At the summit of a mountain, where by theory the air has lost one half of its density, the cold compared with that at the surface of the earth below is most intense. Heat, it is said, always rarefies the air, cold condenses it; and of course there would, from this cause, be a greater density in proportion to the elevation.

It is also settled, that the force of the elasticity of air increases with its density. Of course elasticity acts against the rarer strata, and this elasticity would of itself preserve throughout the whole atmosphere an equilibrium of density. If a quantity of air be artificially condensed, how quickly will it return to its normal state when the force which compresses it is removed! If the air be an elastic fluid, no volume of it can exist without an equal pressure in every direction, — be the volume large or small, a few cubic inches or the whole atmosphere. If any portion of it be rarefied by heat, this will give to its particles greater "force of repulsion," so as to equalize its pressure with that of the surrounding denser air.

Moreover, if gravitation could press down the different strata of the air in the ratio supposed, it could not maintain the pressure, — it could not preserve this difference of density. There is a perfect intermobility of the particles of fluids, and every wind that blows would tend to bring back the equilibrium. It is supposed that winds are caused by the rushing of air from the colder and more condensed portions of the atmosphere to those comparatively more rarefied. An upward current would at once be established to restore an equilibrium, — to undo what gravitation had done.

But the law by which the atmosphere is packed according to barometrical indications, is not the law which regulates the gravitation of fluids. It is a law which acts in this case only. The law, by which one stratum of the atmosphere bears not only its own weight, but that of all the strata over it, is assumed for special purposes. Having discharged its function, it is, ever after, a dead letter. We hear no more of this vertical gravitation, — no more of the bearing down of one stratum upon another. Fluids press equally in all directions. This is acknowledged as a general law. The very action of the atmosphere on the barometer must often be by an upward gravitating power, pressing against the leather at the bottom of the cistern containing the mercury.

The acknowledged principle that the pressure of fluids is equal in all directions, "upwards, downwards, and sideways," would prevent such a compression of the atmosphere as the barometer is supposed to indicate. If a cubic foot of air presses equally by its six sides, it presses the cubic foot above it with the same force as the cubic foot below it. In no way whatever can it be shown, without contradicting all the present theories, that there is a difference of density among the different strata of air. If this difference actually exists, we must give up gravitation with force according to distance, the action of heat in the rarefaction of the air, the elasticity of air in proportion to its density, the diffusiveness of air, the intermobility of the particles of air, and the principle that the pressure of fluids is equal in every direction.

But in fact the varying density of the air has no proof whatever. It may be, or may not be. If it be, it is caused by a principle not embraced in any of the present theories, and it cannot be proved from the range of the mercurial column. We wish to express this idea distinctly. If the air be more dense at the surface of the earth than a mile above it, the increase of density does not arise from gravitation; and, if there is a difference of density at different levels, the barometer cannot indicate it.

We referred, in the preceding chapter, to the oscillations of the mercury at one supposed level; we have occasion to use again the same facts, and to add to them others of the same character.

These oscillations, of which the range is about three inches, while the barometer remains at the same place on the earth's surface, are not accounted for by the present philosophy. Scientific men confess their inability to explain them. Professor Daniell says, "The cause of these oscillations has long been the subject of investigation with philosophers, and the problem in all its generalities is difficult and complicated." "The difficulty," says an article in the London Encyclopedia, "consists in explaining why these oscillations are greater in high latitudes than between the tropics, as well as why they should exceed in all cases the quantities which calculation of different densities of atmosphere might assign." Again: "It is hard to say what can be the cause of the changes of the height of the mercury at the same level, — changes which would be equal to a change of level of two or three thousand feet."

It is supposed by some that different degrees of humidity in the atmosphere might account for these changes. Professor Leslie inclines to this belief, "because all other assigned causes are insufficient." But the barometer usually falls in a damp atmosphere, and vapor of itself has weight, nor does it necessarily rarefy air which contains it. The winds passing over the surface of the earth cannot be the cause; for, as we have already shown, it would require a wind passing over nearly three thousand miles at the rate of sixty miles an hour, to change the height of the mercury half an inch. Attempts have also been made to trace the cause in the vacua occasioned by conflicting winds. In vain; for winds by theory restore equilibrium,—they do not create inequalities, nor could a vacuum exist for days and weeks if it could be formed.

The maximum of oscillation is near the parallel of 45° latitude. The oscillations vary with the time of the year, (being greater in one month, less in another month,) and also with the hour of the day. They also vary with sudden changes of the level of the water, with the height of the tides, and with the position of the moon in relation to the earth. All these periodic and other changes cannot be traced to the state of the atmosphere as the cause. Therefore, if the barometer indicates varying density of the atmosphere, it also indicates some other fact,—the oscillations have some other cause. If it do not indicate the density of the air in one position on the earth, it may not in any, especially as the corresponding variation of density has no proof independent of the change in the mercurial column.

But it may be thought that we have forgotten the fact that, "if a man ascends a high mountain, he will feel an inconvenience in his ears, and other cavities of the body from the dilatation of the enclosed air adjusting itself to the diminished outside pressure." This fact, if it be true, is insufficient to bear the burden of proof of the great principle under discussion. If uncomfortable bodily sensations are experienced, it does not follow that the reason for them is the "dilatation of the inclosed air," or the rarefaction of that outside. There may be other reasons for painful sensations on the ascent of high mountains.

The effect on the animal structure of the air at great heights, we think, proves distinctly that it is not so rarefied as is supposed. The bird soars as lightly from the top of the high mountain as from the plain. The condor, the heaviest of birds, flies without effort at a height of fifteen thousand feet, though supported by air of only half the density of that below. And man, after recovering from the fatigue of ascent, is not always distressed for breath; he does not require lungs of double capacity, nor does he breathe twice as fast to obtain his usual supply of vital air. No one could live comfortably in air of only half its usual density; yet in the highest ascents in balloons we read that "the aeronauts suffered no inconvenience in their respiration or other animal functions." We shall again recur to this part of the subject.

It may also be alleged that the boiling point of water varies according to elevation, because the pressure upon the surface of the water is diminished by the rarefaction of the air; or that, as water boils in vacuo at a very low temperature, so as we ascend and the air is rarefied, water boils with a less degree of heat. The reason why water boils more easily the greater its elevation, is simply its possession of a rotative force proportional to its level of rotation. Beyond a certain elevation water would at once pass into vapor.

We do not expect the minds of others, at this period of the investigation, to accept our explanation of the vaporization of water at different levels. We will, therefore, for the present, admit that the atmosphere is more or less rarefied according to the altitude of its strata,—that the mean weight of the air, from the action of heat or other causes, varies, so as to make a column of it bear on the earth sometimes with a greater, sometimes with a lesser pressure. But even if we admit the gravitation of air, and the varying weight of a given quantity of air, it does not follow that either its normal pressure or its variation of pressure is measured by the stationary or varying weight of the suspended column of mercury. In other words, there is no proof that the weight of the column of mercury is the measure of the weight of the air.

In the first place, when the barometer is constructed, the mercury in the tube does not owe its elevation to the pressure of the air. Let a tube closed at the top be immersed in mercury, and when filled let it be withdrawn upward by the hand; the mercury is raised with the tube above the level of the mercury from which it is taken. The force required to lift it is just the weight of the mercury and tube together. The pressure of the atmosphere on the lower surface of mercury does not in the least degree aid in the elevation, does not diminish the weight,—because the sustaining mercury is pressed by the air, not one iota less of force is required to give the sustained mercury its elevation. The fact is admitted, and is accounted for in this manner:—"The force used in lifting the mercury is needed to elevate the air over the tube; that being done by the applied force, it is the weight of the atmosphere which elevates the mercury in the tube." But this reason is given without reflection; for the air is no heavier over the tube when the tube is filled with mercury than when it is filled with water or air, yet the force applied measures the weight of the fluid within, and the contained weight does not increase or diminish the weight of the column of air over the tube. It is certain, therefore, that when the column of mercury is lifted with the tube, it is the applied force that lifts it, and not the weight of the atmosphere bearing on the surface of the mercury from which it is withdrawn. Of course, the weight of the atmosphere, not capable in the least of aiding the process of elevation, has no efficacy in sustaining the column when raised. One would almost suppose that an air-supported column of mercury would be like the air-supported balloon, and would not press with any weight upon the instrument.

When the tube is elevated beyond what may be termed *the limit of extension*, the mercury is no longer lifted with it. However high the tube be drawn, the surface of the mercury in it remains at a certain distance from the surface of the lower mercury. This distance, at the mean level of the earth's surface, is about thirty inches. It varies according to altitude.

In the usual process of constructing barometers, the tube is filled with mercury, and its lower end immersed in a basin of the same fluid, technically called the cistern. The mercury falls from the top of the tube to the limit of its suspension; and as much as falls from the tube is raised in the cistern. As much mercury is elevated in the cistern as is depressed in the tube; the descent and ascent are exactly equal. In all subsequent changes the fall of mercury is equal to the rise, particle for particle; the flow of mercury back and forth makes the same relative change in the volumes of the tube and cistern mercury. A perfect equilibrium was at first established by the ascent of a quantity of mercury in the cistern exactly equal in amount to that which descended in the tube, and the equilibrium is ever sustained by the same process, without reference to atmospheric pressure.

With this perfectly balanced instrument the atmosphere is supposed to be weighed. It is in one scale, but no test weight is put in the other scale. The air is weighed against nothing. All the ponderable matter hangs on one arm of an evenly balanced lever, and its weight is to be ascertained by the degree of the ascent of the other arm. Place on the outer mercury an additional pound of the fluid, the mercury in the tube will not rise so as to weigh a pound more; add to the density of the air over the outer mercury, the rise in the tube will not correspond to the increased weight of air on or over the cistern. When we added mercury to that in the cistern, the mercury in the tube rose just enough to maintain its former distance from the lower surface. When the tube was first plunged in the cistern, it made no difference how great or how little was the quantity of mercury in the cistern; it makes no difference afterwards whether mercury be added or taken away; the same distance will be preserved between the two surfaces. The rise of the fluid in the tube will not, therefore, measure the weight in and on the cistern.

It will be said, however, that though, when an additional pound of mercury is placed in the cistern, an equivalent pound does not rise in the tube, enough rises inside to correspond with the increased *height* outside, and therefore it weighs a column of mercury resting on the surface of the outside mercury, which is of the same area as itself, and consequently, in weighing the air over the cistern, it weighs a column of air of its own area only.

This idea is a mere statement of facts. The result empirically obtained is exalted into a principle, and one which is at variance with all other principles of the action of fluids. Fill a cylinder with mercury, and from the cylinder let a long tube ascend; subject the mercury in the cylinder to the pressure of a piston, and the rise of the mercury in this tube, that is, the weight of the mercury supported by the pressure of the piston will be measured by the whole pressure of the piston, not by its pressure on a surface of mercury in the cylinder, equal merely to the area of the tube in which the mercury is supported. So if the rise and fall of the mercury in the barometer are proportional to the weight of the air, they are so to the weight of all the air over the cistern; but, as it makes no difference in the action of the barometer whether the surface of the cistern is of two inches area, bearing a column of air of two inches area, or is an ocean of mercury, bearing the atmosphere of a hemisphere, we are compelled to look for some reason besides the weight of the air to account for the fact, that the two surfaces of fluid in the barometer preserve a fixed distance between them. In other words, the variation in the length of the column of mercury is not determined by a corresponding variation in the density of the air.

In the use of mercury to indicate the pressure of steam, or of the condensed air of blowing cylinders, the extent of its rise in the indicating tube corresponds to the extent of the surface of mercury exposed to it. In a steam gauge the area of mercury exposed to the pressure of the steam bears such a proportion to the area of mercury in the indicating tube, that the usual range of force is measured by its rise and fall.

Now, enlarge the surface exposed to the action of the steam, give it an area equal to the area of the piston, the gauge would burst by the increased pressure, and the mercury be thrown with violence in all directions. Take out the barometrical tube from its cistern, and place it in an ocean of mercury, its range would not change, — the two surfaces would still remain at the same distance apart.

Why does the mercury select one column of air of its own area and answer to its pressure, regardless of the pressure of a hundred other columns of the same size and weight which bear upon the cistern? It cannot be from the uniform bore of the tube; for the tube may be unequal, either contracted or enlarged; it may be a cone with its base downward, or a cone with its base upward; the aperture or opening between the tube and cistern may be larger or smaller than the bore of the tube; the relative quantity of the mercury in the cistern may be such, that on bringing the instrument to a lower level it will nearly all be drawn into the tube, and yet will remain suspended, so that in this case, if the mercury is sustained at all by the atmosphere, it is by a column of less area than the bore of the tube pressing on the small orifice between the tube and cistern. There is indeed nothing in the action of the barometer which proves that the mercury in the tube is sustained by a column of air of its own area.

Further; if it were possible that the sustained mercury under any circumstances could select a column of air of its own area, out of a volume of air pressing by elasticity in every direction, and disregard the pressure, weight and elasticity of other columns which act on the surface of the cistern, yet it would be impossible as the barometer is usually constructed. The cistern is closed at top; its bottom is a piece of leather; the instrument is inclosed in a wooden case and suspended perhaps in a close room. It then receives the pressure or weight of its selected column of air forty-five miles high, one quarter of an inch in diameter, which may be twisted and beat about by the winds of heaven, and reaches the instrument through doors or windows, passing through the air in the room, which is subject to expansion and condensation by artificial changes of heat. Yet, not affected by these changes, it penetrates through the openings of the wooden case, acts on the bottom of the leathern support, and still gives exactly the degree of pressure belonging to its area, to the mercury in the tube through an aperture of any size! Is it possible that any thing can be measured by such an instrument, except, perhaps, the elasticity of the air in the apartment in which it stands, which would affect the whole extent of the surface of the mercury on which it acts?

But if this last and only possible action of the air by its elasticity be the cause of the oscillation, why is it that two upper surfaces of mercury, one in the cistern and one in the tube, are an essential condition of the oscillation? A tube in a conoidal form with a piece of leather over its base, would as well indicate the pressure of the atmosphere. The leather would act as well thus placed. It would have the same elasticity. The air pressing on it would act as truly without an upper surface of mercury as with it. The fact that two upper surfaces of the fluid are necessary, abundantly proves that another cause for the action of the barometer is yet to be discovered.

In our examination of the action of the pump and of the syphon, we hope to throw more light on the subject of the elevation and support of fluids by means of a vacuum.

Surely enough has been already said to throw a strong suspicion over the theory which explains barometric action by atmospheric pressure. Indeed, the explanation has often been received with some distrust. A friend, whose strength of mind was never questioned, said, "I have believed the explanation, relying on the authority which gives it, rather than because it satisfies my mind; there is something about the action of the barometer which is not yet understood."

If both the construction of the barometer and the qualities of the atmosphere are such as to throw great doubts over the supposed action of the weight of the air on the mercury, how much are these doubts increased when it is known that the mercury changes its level in the tube, under circumstances which forbid the idea that there have been equivalent changes in the density of the air? To place this matter in a strong light, we will give the following statement from Humboldt's Cosmos: — "The horary oscillations of the barometer between the tropics present two maxima, namely, at 9 or 9¼ A. M., and 10½ or 10¾ P. M.; and two minima, at 4 or 4¼, P. M., and 4 A. M., occurring, therefore, in almost the hottest and coldest hours. Their regularity is so great that, in the daytime especially, the hour may be ascertained from the height of the mercurial column without an error, on the average, of more than fifteen or seventeen minutes. In the torrid zones of the New Continent, on the coasts, as well as at elevations of nearly thirteen thousand feet above the level of the sea, where the mean temperature falls to 44° 6′, I have found the regularity of the ebb and flow of the aerial ocean undisturbed by storms, hurricanes, rain, and earthquakes."

Can we receive the idea conveyed in the latter clause of this quotation? If there is an ebb and flow of the aerial ocean, is it possible that its regularity should remain undisturbed by storms and hurricanes?

If the oscillations of the barometer day after day, month after month, are as regular and uniform as the beating of the pendulum of the clock, these changes come not from the capricious, fitful alterations of the state of the air. The action of the most unrestrained element in nature gives not results the most exact and most regularly continued. If, as Humboldt expressly states, the horary oscillations of the barometer are not affected by changes of atmospheric temperature, neither by heat nor by cold, by snow nor by rain, by storm nor by calm, by the peace of nature nor by the throe of the elements, if the changes go on regularly and periodically on the plains and on the tops of high mountains, surely it is vain to seek in the variations of atmospheric pressure the cause of these oscillations.

Geologists have often noticed that the changes of the level of different portions of the earth are complementary. There is a rise in one place corresponding to the fall in another place; for instance, an elevation taking place in Sweden is responded to by a depression in Iceland. De Beaumont calls this the "mouvement de bascule." Besides these occasional movements resulting in permanent change, we believe that there are periodical fluctuations of the surface of the earth of this compensatory nature, occasioned by the varying distribution of the force by which she moves in her orbit. This fluctuation would enlarge the earth in one diameter, and contract the diameter at right angles with this. Thus there are two daily elevations and depressions corresponding with each other. This gives the two maxima and minima of the barometer.

The elliptical form of the earth's circumference is dependent on her position as being nearer to, or further from the centre of revolution; and the shrinking or expansion is regularly propagated by her rotation. The movement being oscillatory, the maxima will, of course, be somewhat after noon and midnight, corresponding with the minima of the barometer. The extreme would also be greatest within the tropics. Connected with this subject are the horary changes of the horizontal magnetic intensity, to which we shall hereafter allude.

To one other fact only will we at this time advert. It is stated, on good authority, that the barometer is more to be depended on for determining heights, than trigonometrical observations are. The reason given is, that "the refraction of the air prevents accurate measurements of the angle." If we are correct, the reason is obvious; but it is impossible that the density of the atmosphere should, under all disturbing influences, always correspond to elevation with a greater accuracy than the results of geometric calculation.

In our attempt to explain the cause of the action of the barometer, we can, in this place, but imperfectly advance the views that have presented themselves. These views involve peculiar ideas, not only in regard to the force of rotation and revolution, but to the vacuum, to the action of fluids in relation to their own particles, and to their action in relation to other matter; and the ideas, to be presented with any distinctness, must be gradually unfolded as we proceed.

The changes of the mercury result in a longer or shorter column of the fluid in rotation as one mass. In the rotation of any one mass, the force of rotation is unequally diffused through it; the parts, say the upper and the lower, rotate with greater and less force. This inequality of present force increases with the distance of the two surfaces. It increases also with the density of the mass, the distance between the surfaces remaining the same; it increases by a greater degree of rotative velocity, in a mass of the same density and same distance of the two surfaces. In other words, while the ratio of difference remains the same between the different parts of the same body, the actual inequality is increased by increased extension, or density, or velocity.

We have, then, an unequally distributed rotative force present in every mass, and the extent of the inequality is in direct proportion to the difference of the orbit of the upper and lower extremities, to the density of the body, and to the velocity of its rotation.

In solids bound together by cohesion, in fluids so placed that the upper and lower surfaces cannot approach or draw apart, this inequality must and will continue, whatever be their depth, density, or velocity of rotation. But from the law of the diffusion of force, there is a normal difference of diffusion to which all fluid bodies conform when free to adjust their volume in relation to it.

From this principle comes what we term the *limit of extension* of fluids, — a distance between the upper and lower surfaces, which measures the normal inequality of the present force, — an idea which we hope fully to illustrate as we proceed. At present we merely state the fact. The limit is perceived only in fluids so placed that a change can be made in the length of the column. This is the case with the mercury in the barometer. It can shorten or lengthen its column, and it must have the upper and lower surfaces to effect this. It keeps the normal inequality the same under any force of rotation, by extending or lessening the distance between the two surfaces. It rotates with a column which corresponds in length to the force of rotation. If the barometer is made with a fluid of less density than mercury, there is the "limit" for the same cause, but it admits of a greater distance between the two surfaces, under the same force of rotation.

We dismiss the special consideration of the barometer, with a most unhesitating belief that its action is caused by the increase or diminution of the force of rotation, — that in its changes it denotes its level, — that it indicates *directly* the altitude. We believe so from the fact that such would be its action as deduced from principles established in our mind on other grounds. We also believe so, because we find the oscillations always to be such as could be occasioned by changes of level, and often such as could never be produced by varying densities of the atmosphere; and because, if the atmosphere be of varying density, the construction of the barometer is such that it could not be affected by the variation. We solicit a thorough examination of this subject; for, if we are correct, it gives a value to the instrument far greater than it now possesses, its readings being at present confined to atmospheric pressure.

We pass from the barometer, hoping further to elucidate its action by examining the limit of extension in relation to water. The principle is the same in every fluid; but water, for many obvious reasons, most distinctly illustrates the opinions that we would present.

Water, in a tube closed at the top and open at the bottom, on being immersed in water will remain suspended at any height within the limit. But if the top of the tube be opened and air be allowed to enter it, the inclosed water will sink to the level of the water in which the tube is immersed. The exclusion of air from the tube is, therefore, the condition requisite for the water in it to remain above the level of the water in which it stands. This appears of easy explanation. When a volume of water is in contact with air, the surface of this volume of water, if at rest, will be of uniform level. In the two elements, air and water, there is a difference of capacity for force, so that the one, when in contact with the other, invariably retains a fixed degree of force relative to the other. This principle gives uniformity of level to the surface of any volume of water in contact with air; but, if any part of its upper surface is without this contact, its level of rotation does not depend on that which gives level to the parts of the surface of the water which are in contact with the air.

Under the theory of gravitation it is supposed that it is the weight of the atmosphere on the water outside of the tube, and its absence within it, which create the inequality of pressure, resulting in the elevation and support of the water in the tube. This is supposed to be shown by the action of the atmospheric pump, by the syphon, and in fact by all the cases of the elevation of water which exhibit the law of extension. It is considered as occasioned by the vacuum, fluids being suspended because of a vacuum, or because there would be a vacuum if the fluid did not ascend.

The idea is not distinctly advanced at the present day that the action of the fluid is from an attractive power of the vacuum; but the inference necessarily drawn from the facts in connection with present theories is, that nature so abhors a vacuum that action is immediately induced, not only to fill the vacuum formed, but to prevent it from being formed. There is always a supposition of aid from the weight of an outside column of air, of the same area as the column of heavier fluid to be acted upon; but in no case can the action of this external counterpoise be proved, and in many cases it is certain that it does not exist or act. In a succeeding chapter, when speaking of the properties of air, we shall endeavor to illustrate the principle, by which a vacuum artificially created, as it were, draws in matter to fill the void space. In our present view of the subject, we confine ourselves to the illustration of the idea, that the support of a column of fluid within the limit of extension does not and cannot depend on the pressure of a column of air of its own area.

We will suppose a common atmospheric pump, working with perfect nicety, and with the least possible friction. The piston is in contact with the surface of the water to be raised, and on being lifted the water is raised with it. There is no vacuum here; the atmosphere rests on the piston, and the piston on the water, and this mutual relation of atmosphere, piston, and water continues. The piston and the water below it are not in the least relieved of the weight of the aerial column over them; if it pressed on the piston before it was moved, it equally presses when it is moving upward. If the pressure of an equal column of air on the water of the cistern raises the water in the pump, its action prevents a vacuum from being formed. We make for the present the simple statement, that force applied to the water gives it power to rise to a higher level, and the elevation of the piston gives it space, room, freedom to ascend,—the needed force for ascent is supplied, and a channel opened for the action of the force. In all similar cases we shall ever find, that when fluids rise in pipes, it is because of the application of force, with space for the action of the force.

That, in the case of the pump to which we have just alluded, a column of air of the area of the ascending column of water, in other words, atmospheric pressure on the water in the cistern, does not lift up the water in the pump, can be most distinctly proved. We will suppose the piston, which touches the surface of the water, and the elevation of which gives space for the rise of the water, to be securely fastened to the tube, and that the pump, tube, and piston together, be lifted as high as the piston was previously lifted in the pump. The same quantity of water will be raised as before. In the one case, space has been given for the water to rise by elevating the piston; in the other, by elevating the tube with the stationary piston. The force required to raise the tube, valve and water is just the weight of the tube, valve and water. In either case there needs to be supplied the same degree of force, which would be required to raise a stone of the same weight. How then can it be said, that in working the pump, the pressure of the atmosphere is the elevating power, when for every pound of water raised the force necessary to raise a pound is required?

It is said in Ewbank's Hydraulics that the pump will work "if the well be covered with slabs of stone, and coated all over with the best hydraulic cement," thus cutting off from the water of the well all pressure of the atmosphere, except such as is drawn by the action of the piston "through the minute pores." The only necessary connection with the atmosphere is limited to that orifice, which will admit into the well a volume of air equal to the volume of water withdrawn, otherwise there would certainly be a vacuum formed in the well. No connection whatever between the atmosphere and well is needed, if the water elevated runs back through another pipe into the well. The act of pumping is the establishment of a current of water. It has no connection whatever with atmospheric pressure.

A pump would work completely immersed under water, there establishing a connected current, and after the current was formed the upper surface of the water would not be even ruffled, the atmosphere pressing equally on all the parts of the current below.

It is said that the atmosphere presses on the body of every man with a force of some fifteen tons, but because it bears equally in every direction, because it is so exactly equipoised, it is not an object of sensation. We do not feel it. Its force is neutralized, destroyed, so far at least that it induces no perceptible results. This is true, not only in relation to the human body, but to every mass of matter immersed in air. The cistern and the water that it contains, the valves, the water in the tube, and the jet of water issuing from the tube, are as one body; the enfolding atmosphere is wrapped around the whole with equal strain. If the pressure of the air has power to cause the water to ascend through the tube, it has the same power to act in the opposite direction. If it press water in at the bottom of the tube, why does it not press it back with equal force? When the upward current is established in the pump, all the force to create the current must be supplied; the force needed is exactly equivalent to the weight of water elevated. Therefore, the doctrine of equal areas of atmosphere supporting the denser fluid in vacuo has no more application to the raising of water by the pump, than it has to the drawing of water from a well by the pole and bucket. By both mechanical means strength is put forth proportional to the quantity of water to be raised, and to the distance of elevation, the weight of the atmosphere, so far as any practical result is concerned, being neutralized, destroyed, by the affirmed law of its equal pressure in every direction.

12*

But for the skeptical we will mention a fact which perfectly demonstrates that the action of the pump has no reference to the pressure of the atmosphere. By accident, in Seville, in 1766, it was discovered that water could be raised by the pump fifty feet and more, if a small opening was made for the admission of air into the tube of the pump. This circumstance excited great attention at the time, and was supposed to prove that water could be raised by atmospheric pressure much higher than thirty-three feet.

In the Journal of the Franklin Institute for May, 1837, it is stated with reference to the action of pumps worked by steam, that a little air is sometimes admitted into the pump pipes, which "makes the pump work more lively in consequence of the spring it gives to the column of water." And in the same paper, Mr. Perkins states, that "forty years before an attempt was made to impose upon him a pump which raised water by atmospheric pressure one hundred feet; but he detected a small pin hole in the pipe through which the air was admitted."

These facts are fatal to the theory that atmospheric pressure causes the elevation of water in pumps; for there may be the pressure on the ascending as well as on the impelling column. No matter how little this pressure is, though the fact was shown in Ewbank's well, closed by hydraulic cement, that the pressure through small apertures is as strong as through large,— if the pressure within was but an hundredth part of that without, what there was of it would balance proportionally the external pressure, and to this extent too it would impair the vacuum; but instead of this it gives new power of ascent to the water, more than doubling the possible extent of its elevation. To account for this we are told, "It was ascertained, on investigation of these facts, that the air on entering the pipe mixed with the water, which therefore, instead of being carried up in an unbroken column, was raised in disjointed portions or in the form of thick rain. This mixture being much lighter than water could be supported by the atmosphere, and by proportioning the quantity of air to be admitted, a column of the compound fluid may be elevated one or two hundred feet by the atmosphere." In this explanation the fact seems to be forgotten, that it is not the *weight* of the column of water which determines the altitude to which it can be raised. Increase the force, and ten thousand pounds may be lifted as readily as ten pounds. It is the *height* of water, not the weight, which attained, it will no longer rise; and the main fact was quite forgotten, that atmospheric pressure was at work as well within as without the tube.

Our views can be presented more distinctly by reference to the action of the syphon. In the pump the current is established by animal force, or by the power of wind, steam, &c. applied to the piston; in the syphon the same end is obtained by the presence of a longer column of water in one leg than in the other. To use the common language, gravitation acting on the longer column with greater force than on the shorter, the equipoise is destroyed, and the water falls in the one, rises in the other, thus establishing a continuous flow. It is the force of descent of one body drawing up another body with which it is connected, as by a rope passing over a wheel. The strength of the flow is therefore determined by the difference in height of the two columns, the force gained by the descent of one being more than sufficient for the elevation of the other. The flow has no reference to the attraction of gravitation; for by different diameters of the two columns, the quantity or weight of water may be greatest in the shorter leg.

We will suppose a syphon, the shorter leg of which shall be ten feet high, containing a hundred pounds of water, its longer leg of eleven feet containing five pounds of water. The water from the longer leg is discharging itself into the air in a continuous stream. What supports the one hundred pounds of water, which, by theory, is drawn towards the earth by a force equal to its weight, while the water in the longer leg is drawn down by a force of only five pounds? Certainly it cannot be atmospheric pressure on the larger tube; for there is also atmospheric pressure at the discharging orifice of the smaller tube. The attraction of the hundred pounds of water, if it be attracted at all, should be sufficient to draw up the five pounds and to draw in the air after it. Of course the one hundred pounds does not feel the force of attraction.

The action of the syphon is therefore but the fall of water through a tube arched upward, and the force of the fall is measured by the difference of altitude between the two levels, — that from which it is drawn and that into which it issues. In the pump it appears as if a vacuum were created, as if the current were established by the weight of the air pressing on the water in the well, and not on that in the tube. In the syphon there is not the most remote indication that a vacuum is formed, and the pressure of the atmosphere is exactly the same on the water which enters the pipe, and on that which leaves it. There is no vacuum; for the syphon is kept full of water, and this creates a plenum as much as if it were filled with lead; and with the shorter leg immersed in water, the surface of which is at a higher level than that in which the longer leg is immersed, if there be any difference, the atmosphere presses more heavily at the discharging than at the receiving orifice. There can be no pretence of atmospheric action on a current of water flowing from a higher to a lower level, through a tube arched upwards. The air in this case has no more to do with the current than if it flowed through a pipe bent downward, or curved laterally, — no more than if the water ran down through a straight tube, or without any tube through an orifice in the side of the containing vessel. The action of the syphon is imputed to atmospheric pressure, because the limit to which water can rise in it is the limit, beyond which water cannot be raised in the cases in which atmospheric pressure is supposed to be the cause of the elevation. It is true that if the water at the top of the syphon should part, and it should descend in each column, there would be a vacuum ; but, there being no gravitation, it will not part till beyond the limit of extension. It then parts through the action of the law which regulates the limit. But because, if the syphon does not establish its current, and the water falls in both the legs, there is a vacuum, it does not follow that a vacuum or the dread of one establishes the current when it flows. While the current is flowing there is, if any, an equal pressure at both ends, and from them up through the whole length of the pipe.

The vacuum is formed, if the syphon be raised above the limit. If the instrument be extended higher, the water does not rise ; but remains elevated at an equal height in both legs, measuring from the respective levels of the surfaces of the water in which they are immersed. It is supposed, that each suspended column is counterpoised by an equal weight of atmosphere, that is, by a column of air of the same area as the suspended column of water. Of course, this being the case, any variation of the limit marks an equivalent change in atmospheric weight.

Without going into the question at this time, of the possible efficacy of the air in sustaining water in such cases, which may well be doubted, as we have seen that it has not the most remote agency in giving it its elevation, we feel confident that the weight of the suspended column does not weigh a column of air of its own area.

If water be subjected to pressure in a cylinder by a piston acting on its upper surface, the rise of water issuing or sustained by the pressure is measured by the whole amount of the force of the piston, as in the case of a forcing-pump. If the atmosphere presses on the surface of water in a pond, or in a small vessel, the pressure is measurable by the force acting on the whole surface. A pond a mile in area bears the pressure of a column of air a mile in area; and the water in a small vessel bears the pressure of the column over it. On this principle only can water be acted on by the air, and any variations of pressure arising from atmospheric changes would express the variations of the weight of the whole volume of air over the pressed surface.

Thus far we have considered the limit of extension of fluids, and appear to find that this limit is not determined by the gravitation of the fluid, since the quantity sustained under the same circumstances may be one pound, or one thousand pounds; nor by the gravitation of the air on the exposed surface of the fluid, since this may be one inch or a thousand miles in area; nor by any virtue or efficacy of the vacuum, since it may be equally filled, saturated, and thus destroyed by matter of its own volume, whether the matter drawn to fill it is light as air or dense as mercury. The limit is therefore not caused or determined by the gravitating power.

We advance to a farther consideration, — to create the

limit, it is not necessary that the fluid be confined in pipes or vessels. Water out of pipes, unconfined, in the free air, obeys the law. The waters of the ocean respond to this limit. Water-spouts, with mingled air, may raise a broken column to a higher range; the waves may dash themselves farther up on a rocky coast. Wave may be formed on wave, a new limit of extension may perhaps be formed upon the original limit, an ocean wave upon a tidal wave; but recorded events always give a height of the rise of floods, which denotes the bound assigned by nature to the heaping up of the waters.

It is difficult to measure the height of a wave, and impossible to separate exactly the unbroken water from the crest which is mingled with air. We know, however, that the rise of waves is less than thirty-four feet in the most violent storm. One measurement taken in a long-continued heavy gale gave thirty-two and a half feet from elevation to depression; on striking a wall of rock, the unbroken column ascended sixty-five feet from the level of the ocean, dashing its spray much higher. The extent of the rise of the tide in the bays of Fundy and of St. Malo is said to be sixty feet, and the reference to heights of which the simple limit forms the root or multiple, in the statements of the rise of floods, is so frequent as to indicate some principle. It is certain, however, that the waves of the ocean have a fixed limit corresponding to the limit of the extension of water in tubes.

The isolated fact of the limit is thus advanced to an universal principle; for it has never been imagined that an unequal atmospheric pressure causes the rise of the waves. Whenever a fluid is so placed that its degree of extension upwards and downwards can be changed or modified, when it is free to assume its own position, it obeys the law of limit. To give it a greater extension as one mass, it must be restrained or confined. Its spread or dilation in breadth must be prevented, either artificially as in tubes, or naturally as in a containing basin. Its base must be cut off, — it must be separated from other volumes of the same fluid. Water can be elevated to any height by the lifting pump, or by the application in any manner of force beneath the volume of water to be raised. But if the force be applied from above, as in the atmospheric pump, or in any manner to a body of water remaining in connection with other water, the force applied is conveyed away from the special volume to the whole volume; a portion therefore cannot be elevated. It must first be separated, that the force applied may act upon it. It will then rise; but before the separation the force applied is communicated to the whole volume. It is thus that separated volumes may be lifted to any height, if sufficient force be used; and this explains the fact before alluded to, that, when air was admitted into the pump by a perforation in the tube, water was lifted beyond the limit, since the air separated the one volume into detached portions.

The limit of extension is therefore the degree to which fluids can be drawn upward in an unbroken column. It is the limit of the inequality in the distribution of the force of rotation, which will remain when the excess can be transferred to another portion of the same volume. It is the limit of elasticity of form; for while a solid body ever retains one extension, — no one part of it being susceptible of elevation independently of the other parts of the mass, to raise one part requiring force sufficient to raise the whole, — portions of fluids can be drawn up, force, to a limited degree, confining itself to the part. Thus, a column of water can be extended from the volume; but beyond this limit force no longer confines itself to the part to which it is applied, but diffuses itself through the whole volume. Therefore, the sea knows its place; the level of the waters composing so large a relative portion of the globe is preserved, not with a dead uniformity, not with a rigid flatness of surface, but with constant changes, limited undulations; and the ocean, free to move, yet under the law of extension, is as securely bound, as are the rocky cliffs against which its rolling surface strikes.

The more our attention is directed to the unfathomable deep, the more wonderful appears the action of the laws which control it. We are taught to regard the vaporization of the surface of the ocean as a means by which the equilibrium of heat is preserved, and by which the waters circulate from the sea to the clouds, from the clouds to the thirsty earth, and again collecting in swelling streams return back to the ocean in a never-ceasing flow. How great the quantity of fluid thus in constant change, hundreds of millions of tons rising in one day from the comparatively small extent of the Mediterranean Sea, the quantity ever rising from the entire surface of the waters of the globe, from the frozen polar seas, as well as from the tepid oceans of the torrid zones, being beyond all estimate!

This vaporization not only waters the earth, not only restores the equilibrium of temperature, these effects being as it were the incidental mercy of the law of nature; but it is the preservation of an equilibrium of *force*. The ocean exhausts itself of the strength not necessary for its level of motion, the vapor draws off the excess of energy, and the waters preserve their equable flow. The risen vapor, again condensing, slowly imparts a quickening impulse to the sluggishly moving air, or rapidly throws out the force in the lightning's flash, giving to some portion of the earth the needed impulse. For this reason is there greater condensation, more copious rain, in high lands and mountainous regions, which from their elevation require the added force.

No one can contemplate the extent of the oceans of the globe, — their vast surface compared with the land, — without being assured that this unequal distribution is not from accident, — without the belief that there is connected with it the furtherance of some great design. And when we consider its relations to force, its strong conducting power, we see a reason for the ocean's vastness, and comprehend its action in the economy of nature. There is a reason too for the preponderance of water in the southern over the northern hemisphere, which may yet be discovered by man; for, though land and water seem confusedly mingled as if without law, not by accident came the position of the smallest island, nor the bounds of the paltriest lake.

Then again, the waters of the ocean have a molecular action, increasing in intensity as they deepen. The experiments of Scoresby, — the well-known fact of the breaking of a vessel containing air when let down to a certain depth, — the bursting of a tight, full cask, by the insertion of a long slender tube filled with water, — the "*spring*" of water, of which mechanics take advantage in their management of water-power, by narrowing the aperture through which the water flows, — the effect of conical ajutages, by which the flow of water from a discharging orifice is greatly increased from the transfer of the molecular action into the progressive motion of the moving jet, — all these facts together prove that there is an increase of the atomic action of water proportional to the distance from the surface, and that this change is not in the least dependent on a force of gravitation; for water is almost incompressible, and, were it compressed, the increase of molecular action is altogether too rapid for the rate of increase of gravitation, according to the distance from the centre.

Thus the waters receive and retain a greater force than is needed for rotation, resolving it into molecular action, the ocean becoming, as it were, the storehouse of force, from which it may be gradually imparted according as it is needed for the motions of the various parts of the earth. It is by this provision that force divides or diffuses itself equally throughout the volume of the fluid; for, as stratum after stratum requires the less for rotation, the unused force is employed in molecular motion. It is by this equal diffusion that the movable element preserves its continuity; and, without the bonds of cohesion to hold its particles together, it moves with the earth as one mass, no portion ever raising itself beyond the limit, or floating in the atmosphere in detached masses. Without this provision, the air would not be divided from the waters, element would not be separated from element, the waters could not be gathered together into one place.

It appears evident, then, that there is in water and other fluids a molecular action, which increases in intensity according to the depth, or distance from the surface. On the other hand, the rotative force increases as it ascends from the depths, being greatest at the surface. One therefore is the complement of the other, each being capable of resolution into the other. Therefore, force is ever equally diffused through the same volume, the aggregate of force being evenly distributed through each body of water, whatever its depth or extent. It is therefore equal throughout the space occupied by the water.

How far extends this principle of the equal distribution of force relative to space? Is it universal, in solids and fluids, from the densest matter to that which makes the nearest approach to a vacuum? Do accelerated and retarded motion take place from the adjustments of force, as in different degrees it acts in the atoms and in the mass, in the molecular and the progressive motion? Is it a normal adjustment which gives harmony of movement to the spheres? Is motion irrespective of mass? Has the rarer medium of the atmosphere, for instance, an increase of molecular action to compensate, as it were, for its less quantity of progressive motion in a given space? This indeed opens a wide range of thought; but we pause in the inquiry. May we not hope that in time to come the intellect of man may be able to penetrate these mysteries, and to find in the now complicated phenomena of motion that simplicity and comprehensibility, which belong to every domain of nature? If force be equally diffused through space, and we extend the idea of the equality of its diffusion to limitless space, the human mind almost shrinks back at the contemplation of this expanse of energy. There is but one step farther which it can take, going on from boundless force to Him from whom it is the emanation.

To return: there is then by fixed laws a permanence given to the free-moving elements. There is a defined limit of extension, measured by the density of the fluid, and by the velocity of its motion. This is determined by the degree of the inequality of force which will be retained by any volume of fluid, so placed that it can transfer the added force. Broken columns of water may be elevated to any range of rotation; but water resting on water as its base, refuses to extend itself, or to be drawn out above its normal range. Beyond this, force will not so adjust itself as to keep it one mass; if you apply force, the water receives it, but will not retain it. The raging storm passing over the wide ocean, ever imparting force, gives to the water a definite range of altitude only, while the excess is silently transferred to the deeps below. Urge it further and further, it still refuses to rise, and, if saturated, if it can retain no more in the intensity of molecular action, the water at the surface changes into vapor and floats upward in harmless cloud-wreaths.

"And God said, I will establish my covenant with you, neither shall all flesh be cut off any more by the waters of a flood, neither shall there any more be a flood to destroy the earth. I do set my bow in the clouds, and it shall be for a token of a covenant between me and the earth."

The bow in the cloud has been by many looked upon as a figure of speech only, as a beautiful metaphor, possessing no more significance than the clear sunshine after the falling rain. It has not been connected in men's minds with any law of nature, which determines the limit of the rise of the waters. Yet it is the expression of the law by which the floods keep their place. It connects science with the written Word. Understood, the rainbow shows the harmony of the creation with the voice of God. It practically as well as spiritually records the covenant.

The water ascends from its immense ocean bed only as the vapor of the clouds. It cannot rise to deluge the earth. The flowing stands firm, the free moving element is bound in chains; for though intense force, in its might, may rush across the rolling waves, it will not impart to them the power to rise above a fixed limit, except as vapor, and on the clouds the covenant is written in the vivid colors of the rays of the sun.

Devoutly thankful should we be when reason thus re-echoes the voice of God, — when science, in her feeble utterance, repeats the declarations of Scripture.

We cannot, with the thought of this connection, look upon the bow in the clouds but with increased delight, and with a strengthened reverence for the written Word, — with somewhat of the feeling which prompted the exclamation of Coleridge, "What a mine of undiscovered treasures, what a new world of power and truth, does the Bible promise to our future meditation, when in some gracious moment, one solitary text of all its inspired contents dawns upon us in the pure untroubled brightness of an idea, which even as the light, its material symbol, reflects itself from a thousand surfaces!"

CHAPTER VII.

"THERE ARE INDEED MANY THINGS IN THE FRAME OF NATURE, WHICH WE CANNOT REACH TO THE REASONS OF, THEY BEING MADE BY A KNOWLEDGE FAR SUPERIOR AND TRANSCENDENT TO THAT OF OURS, AND OUR EXPERIENCE AND RATIOCINATION BUT SLOWLY DISCOVERING THE CONTRIVANCES OF PROVIDENCE THEREIN.

"NO MAN EVER WAS OR CAN BE DECEIVED IN TAKING THAT FOR A TRUTH WHICH HE CLEARLY AND DISTINCTLY APPREHENDS, BUT ONLY IN ASSENTING TO THINGS NOT CLEARLY APPREHENDED BY HIM."—*Cudworth.*

WE continue our examination of the action of force upon fluids. Our views bring us to the conclusion that there is no weight or pressure of the particles of water against each other, against a containing vessel, or against a foreign body immersed, other than arises from the force of molecular action, which manifests itself under certain circumstances. We do not believe that gravitation presses together the particles of water, or presses them against an immersed body.

This proposition at first sight appears to contradict the generally received opinions on the subject. It is not so. Though expressed in different language, our proposition is in harmony with established views. The law of gravitation is modified in its application to fluids. It is said, "the particles of fluids gravitate independently of each other, not only downwards, but upwards and sideways." "The pressure of fluids is equal in all directions, being founded on the complete intermobility of the particles of the fluid, and on the equal propagation of pressure in every direction." "When a fluid is at rest the pressure will be nothing, because the opposite dead pressures will be equal." These quotations are from the highest authority, and completely prove the proposition with which this chapter commences, so far at least as relates to any result from the pressure.

Admitting, however, the theory that the atoms of water do press against each other, and against an immersed body, the pressure, if it is equal in every direction, cannot proceed from an attraction of gravitation drawing towards the centre of the earth. Why this equal pressure, of the existence of which there can be no proof, is attributed to a force acting downward only, does not appear.

A consideration of the weight of a mass of matter rigidly at one level, or of the pressure of one mass against another, or of the mutual pressure of their atoms when relatively at perfect rest, involves what may be called transcendental dynamics. It treats of a *tendency* to fall, a tendency of the particles of matter to move against each other, which is followed by no action. The tendency to fall, of a body secured so rigidly that it cannot be moved, can never be the subject of experiment. We can measure force only by resultant motion. There is no force of descent unless there is a descent. A body secured at one level at perfect rest, would feel heavy only by the upward pressure of the hand against it. Suppose a body which by its fall presses a spiral spring, the weight of the body is its force of descent upon the spring, which, if it have not force of elasticity enough to throw it back, will remain compressed. The compression is a change in the position and arrangement of the atoms composing the spring. This comparison will indicate the reason of the feeling of pressure in the hand supporting any mass of matter.

We will suppose a vessel ten feet high containing water. From an aperture near the upper surface the water will run with but little force, falling down almost perpendicularly; from an aperture in the middle it spouts with considerable power; and from one at the bottom the flow is with greatly increased energy. The jet is horizontal in proportion to the distance of the orifice from the upper surface; its force is in proportion to the descent of the water inside the vessel. There is little or no force of descent in the water flowing from the upper orifice, — the whole force of descent is in that flowing from the lower opening. It is the force of descent which impels the water, not abstract weight or tendency to descend; for a pound of water at the surface has as much desire to go down, is as much attracted, as a pound at the bottom of the vessel, that is, neither is attracted to the earth. Matter itself is inert, without power to move or tendency to move. It obeys the law of force, and if absolutely at rest it is without force; nothing can be affirmed of it, but that it is, — the fact that it exists in space. We proceed to more practical illustrations of the pressure and equilibrium of fluids.

We will describe from nature. We were leaning on the rail of a bridge, a few rods above which rose a dam about ten feet in height, which held back a pond of water extending beyond the dam for at least a mile. The dam was perhaps two hundred feet long, it was old and dilapidated, and from between the logs there were at every height countless jets of water. On the top of the dam was a flash board, — a plank placed edgewise, slightly secured, to increase the height of the dam. At one place this flash board had given way, and there the water poured over the dam in an unbroken sheet. We will look at this dam with its many jets of water, the broad sheet tumbling over a part of it, to understand the nature of the pressure of water. It will give us the truth, if observed with minds free from all predetermined theories.

In the first place, we see that the water does not press *equally* in all directions, upward, downward and sideways; for the lower jets of water issue with much more force than the upper jets. There is a gradual diminution of pressure from the lowest to the highest; the top of the water falling over the dam does not press at all, for the surface retreats, forming a curve. The pressure increases from the surface to the bottom. Fluids do not gravitate equally in every direction, or the force of gravitation would have thrown out equal jets from the top to the bottom. Perhaps the water gravitates downward only. That cannot be the case; for then there would be no lateral pressure, but at each jet the water would flow nearly without force. One other thing is certain, that this great volume of water, two hundred feet broad and a mile long, does not press laterally with a weight in proportion to the mass of the heaped up water. So weak an inclosure could hardly bear a ton's weight. There is lateral force then only in that water which runs over the dam, or escapes through the apertures. That which is at rest can have no pressure; for, if it had, the weak barrier would be overthrown.

Suppose a stone weighing one hundred pounds, pressed equally one very side by springs. With the pressure of these springs it weighs just one hundred pounds as it did without them; for the action of one spring neutralizes the pressure of the opposite spring, so that the "gravity" of the stone is not changed. We will take away the springs and weigh this stone under water, which by theory presses on it as did the springs, equally in every direction. It has lost weight,—therefore water does not press equally in all directions on the stone. Nor does the water press downward only; for in this case the stone would weigh more in water, "having to bear not only its own share of attractive force, but the weight of all the strata (of water) over it."

How, then, can we ascertain the specific gravity of the stone on both or either of the two laws of the gravitation of fluids? An equal pressure on all sides would be "dead pressure," and the stone would weigh the same in water as in air. A vertical pressure would increase its weight.

We once heard two men disputing on this subject, one affirming that a stone did weigh less in water, because he had often tried it; the other affirming that it could not weigh less, "for you see," said he, "the stone is as much drawn down by gravity in the water as out of it, and the water cannot lift any, when there is more water over it bearing down than under it lifting up."

In fact, on either or all of the principles affirmed to account for the pressure, the change in the gravity of the stone cannot be explained. It is true that the stone displaces a quantity of water equal to its bulk, but this fact neither increases nor diminishes the force with which the earth attracts it. What is the action of the surrounding water? How does it press on the stone? Downward only? Surely not; "fluids press equally in every direction." Equally on all sides? Then the "opposite dead pressures" must entirely neutralize each other, and have no effect whatever. The decreased weight of the stone can be accounted for only on some new principle.

We will bring the pressure of water to an infallible test, that of our sense of touch, — our honest perceptions which favor no theory whatever. Immerse the hand in water; in the act of immersion pressure is felt, for the water is moved by the introduction of the hand, but when it is at rest in the water not the least pressure is experienced. Place the hand near a small orifice from which water is issuing, and pressure will be felt because there is a motion of the water. Stop the orifice with the palm of the hand, and there is no pressure; for the water has ceased to flow.

We now pass to the consideration of that action of fluids which is called the hydrostatic paradox, — an increase of pressure which is not in proportion to the force applied, but is in proportion to the surface of the pressing fluid. A multiplication of pressure is apparently produced, — an effect greater than the cause, — by some supposed mysterious "principle of fluidity" applied force in some cases increases its power according to the extent of the surface against which it acts.

It is said that "the true laws of the equilibrium of fluids were discovered by Archimedes, and rediscovered by Galileo and Stevinus, the intermediate time having been occupied by a vagueness and confusion of thought on physical subjects." The doctrine is that, as a fluid is a body, the particles of which have a perfect intermobility, "therefore all pressure exerted on one part is transferred to all other parts." Stevinus deduced from this principle, that the pressure on the bottom of a vessel filled with fluid, may be greater than the whole weight of the fluid. And it is also deduced, that the pressure may diverge, and may be multiplied in every direction. Pascal shows, in his Treatise on the Equilibrium of Fluids, "that a fluid inclosed in a vessel necessarily presses equally in all directions by imagining two pistons, or sliding plugs, applied at different parts, the surface of one being a hundred times greater than the other; it is clear that the force of one man acting at the first piston, will balance the force of one hundred men acting at the other." And therefore he concludes that, "a vessel full of water is a machine which will multiply force to any degree we choose."

The apparent multiplication of pressure is indeed a difficult subject to grasp or comprehend. The explanations given of the facts are far from being clear. This seems to be admitted; for, as Whewell remarks, "there is a difficulty of holding fast this idea of fluidity. Even at this day, men of great talents not unfamiliar with the subject, sometimes admit in their reasonings an oversight or fallacy with regard to this point. The importance of the idea when clearly apprehended and securely held may be judged of from this, that the whole science of hydrostatics, in its most modern form, is only the development of this idea." But has all "vagueness and confusion of thought" on this subject passed away? Is it now clearly understood how there can be an equal pressure in every direction, and at the same time the pressure be increased according to depth, and again multiplied according to surface?

The facts which are supposed to prove the multiplication of pressure in fluids are, the elevation of a great weight by a less weight of water, as in the hydrostatic paradox,—the bursting of a full cask by the pressure of a very small quantity of water poured into a long tube inserted in it,—and the floating of bodies in water of less weight than the mass that is buoyed up.

The hydrostatic paradox, as it is called, may be exhibited

in the following manner: Insert a long slender tube into a cylinder with a movable piston; pour water into the tube, and the piston will be elevated by it; if the tube holds only a pound of water, and has an area only one hundredth part as great as that of the piston, the pound of water poured into the tube will elevate a hundred pounds placed on the piston. The same action is shown in the Bramah press, by which a great pressure is produced by gradual increments of force, from water driven into a large cylinder by a small forcing-pump.

The explanation which is usually given of these facts is, "that as the pressure of the particles of fluid is equally distributed among themselves, so external force or pressure is distributed in the same way." . . "In the hydrostatic press an immense accumulation of force is brought to bear upon a particular point, by pressure applied by a small column of water *reacting* upon a large mass placed on the surface of the piston." This explanation is hardly more than a recital of the facts; how or why the small column of water should react with a force proportional to the surface against which it is applied is not in the least explained.

The fact that the weight or pressure of one pound of water may be made to produce a pressure equal to that of a hundred or a thousand pounds, is in reality no more paradoxical than that one pound on the long arm of a lever should balance a greater weight on the short arm. Its action is similar to that of other mechanical powers. There is no increase or multiplication of the applied force. The descent of every pound of water gives a spare force of descent sufficient for the elevation of a pound weight. It will raise a pound to the same height from which it fell, or one hundred pounds $\frac{1}{100}$ of the same height.

But the force with which the weight is raised depends also in part on the area of the surface of the piston. It is equivalent to the force of a descending column with an area equal to that of the piston. That is, the weight is raised with a force as great as if the area of the descending column were equal to its own, while the extent of the elevation is proportional to the volume of the water which descends. The extent of the area of the descending column does not change the force of ascent of the column acted upon ; whether it is of one inch or ten inches area, the effect produced is the same. This constitutes the paradox, and it is this fact which requires explanation.

We consider this fact as strong proof of the correctness of our hypothesis regarding the nature of the molecular action of fluids. Every volume of water has its molecular force equally diffused through it. This force must of course be the same in the cylinder as in the tube, because from their connection they form one volume, while without the connection the force would be different in the two columns, on account of their differing depths. The molecular force due to the greater depth of the long column is diffused through the less depth of the water in the cylinder. And, being equally diffused through this body of water, its action is of course in proportion to the area of the water. The abnormal molecular action of the shorter column comes from the equal diffusion of the normal force of the longer column, and, being abnormal in the shorter, it is converted into progressive motion. Of course, the force of the progressive motion acts upon the whole surface of the piston against which it is applied.

We will endeavor further to explain this idea. In the long slender tube the force of molecular action is in proportion to the depth of the water that it contains, but it equally diffuses itself in the water of the cylinder; this diffusion gives an abnormal force for the depth in the cylinder, and it is therefore changed into progressive motion, which is the force for elevation. In other words, it is the height of the column which determines the force of the pressure, and the quantity of water falling which determines the extent of the elevation. The reason that the force of elevation is in proportion to the area of the surface of the water which lifts the piston, is that the force of molecular action which is changed into progressive motion is determined by this area. The idea may be illustrated by comparing the molecular force to spiral springs placed beneath the piston, the number of the springs being determined by the extent of the surface against which they are placed; of course, the larger the surface, the greater the number of the springs in action under it. Hence, "the force of one man may be made to balance that of a hundred men;" for the "reaction" is the joint force of the one hundred and one men equally distributed. And hence it is, that "not only does the pressure diverge, but is in all directions exactly equal, an equal extent of the fluid being taken."

It has also been deduced from this idea, that, if any fluid in a cylinder be pressed by a piston, the force which moves the piston bears with the same power on every part of the surface of the cylinder. For instance, if the area of the piston is one foot, and it is acted upon with a force of one hundred pounds, and the cylinder has an area of eight feet, the one hundred pounds of pressure would be multiplied into eight hundred pounds, to give the one hundred pounds pressure to each foot of the cylinder. But it is not so. The one hundred pounds force is not in the least increased by the size of the cylinder, and is equally distributed over the whole surface. The mistake has arisen from the fact, that, if the cylinder is opened at any part, there will the whole force concentrate itself; and because it can always be brought to any one point, it has been supposed that it acts with its whole force at every point at once. The molecular force is converted into progressive motion only where it acts against that which is movable; it exerts itself where there is room for motion. When confined, the force is equally distributed throughout the containing vessel; when suffered to escape, its whole strength is transferred to the place where there is room for motion. Tap the cylinder at any part of its surface, and thence the fluid issues with the force which theory assigns. Let the boiler of a steam-engine burst, and at the rent will issue the force which theory has made present at every part. The belief, however, may have had a good practical result. The mechanic, on its assumption, gives to his cylinders a strength in every part capable of sustaining the whole pressure. The yielding of any part of the containing vessel concentrates the force at this point of comparative weakness, where the molecular action of the fluid can change itself into progressive or consentaneous motion. The force applied is neither increased nor diminished by the extent of surface against which it acts. It is an invariable quantity, and is equally distributed throughout the containing vessel.

It has been a matter of great surprise to practical mechanics, that steam-boilers have continued to work safely, when worn so thin that it appeared impossible that they should bear the pressure which theory assigned. It was recently mentioned to me by a friend, a gentleman of great practical experience, that he had just examined the boilers of a very powerful engine, which had been taken out for repairs, and was exceedingly astonished to find them so worn, so much corroded, that in some places he could indent them with a slight pressure.

The force of molecular action, however, will not account for the buoyancy or floating of bodies. The water around them is of equable depth; there is no confinement or unequal pressure to produce an increased molecular action in any one part of the volume of water. The molecular force exists in its normal state as the counterpart or complement of the rotative force. Nor is buoyancy produced by the gravitating power; for, as before observed, bodies will float in water of less weight than themselves. Place one cup within another so that their surfaces will be nearly in contact; pour between them a little water, not one half as much as the weight of the inner cup, and the cup will be lifted, and will float. It is a perfectly established fact, that ships do not require to float them a quantity of water equal in weight to themselves. The water in repairing-docks, for instance, may be far less in weight than the aggregate weight of the floating ships. This fact is in direct opposition to the law of gravitation. It cannot be reconciled with it; for the greater weight should be drawn down, and should displace the less weight. Even in the paradox the gain of force is apparent only; it is like the action of the lever; a larger weight is raised only in proportion to the change of level of a smaller weight, the quantity of motion being the same in both. By no mechanical contrivance, whether it act on fluids or on solids, can there be a positive gain of force. A pound cannot, by its pressure or by its fall, sustain at or lift to its own level more than a pound.

If there is a gravitating power, it must act on the same mass, at the same distance from the centre of attraction, with uniform intensity. A cubic foot of water will invariably be acted upon with the same power. In fact, a measure of water forms the standard of weight. Knowing therefore the weight of a body to be lifted up, or to be held suspended, the quantity of water, which would be drawn down with sufficient force to bury up the mass immersed in it, could be exactly calculated. The weight of the water would measure its pressure, — its effective force, — its elevating capacity. The extent of surface against which it acted could not alter its efficacy. If, therefore, in the phenomena of buoyancy, the heavier mass goes up or remains up, and the lighter goes down or remains down, gravitation, acting either directly or indirectly, cannot be the cause or principle by which buoyancy is produced.

It is well known that the displacement by any floating body, a ship, for instance, of a quantity of water equal to its own weight, is the necessary condition of its buoyancy. But the displacement of the water is not the cause of the buoyancy. There is no virtue or efficacy in the place from which the water is excluded. The pressure which gives support must be outside; it must be in the water around the ship. If it is supported by the gravitation of the water, by its pressure against the ship, the question is, on what principle does the pressure act? The extent of the pressure, as we have seen, is not determined by the quantity of the surrounding water. That may be more or less. Nor by the depth of the space in water occupied by the floating body; for it may be wide and shoal, or narrow and deep, and yet the quantity of water displaced be the same. Nor is the pressure determined by the extent of surface exposed to it. Bodies of different surfaces may yet exclude the same quantity of water. If the pressure, therefore, is determined neither by the quantity of water, nor by its depth, nor by the extent of surface presented to it, gravitation is not the cause of buoyancy.

The fact of buoyancy is, we think, susceptible of explanation on our principles. We will suppose that a vessel contains one hundred pounds of water, and that there floats in it a block of wood weighing two hundred pounds. We will mark the level of the water. Take out the wood and pour in two hundred pounds of water, and the level remains as before; the water added, or rather two hundred pounds of water, occupies the place of the wood. It acts and is acted upon just as the wood acted and was acted upon. It is in the place of the wood; that which supported wood now supports water. Enlarge the vessel, and pour in water enough to make the level the same as before. The two hundred pounds of water or of wood retain the same place with a greater thickness of the strata at the sides. And reverse it,—narrow the vessel, and take out water,—the wood floats as before. The support neither of the water nor of the wood was owing to the quantity of water around it. The floating body then does not remain suspended from the weight of the water in which it floats. There is no tendency in the water to crowd it out of its place, nor any to hold it in its place. A floating body has no tendency to sink or to rise. It is not attracted downwards. It rests self-poised. It preserves its level because it has the rotative force of its level; it floats because the rotative energy is equally diffused through itself and the water. A ship has no weight on the water; she does not bear upon it; the hand placed between her and the water would not be in the least compressed.

She is not attracted downward by gravitation, nor is the water attracted upward against her. But if she descend from her level, she has force of descent, as have all falling bodies. She then bears upon the water, moving it by her force of descent. The hand under her would then feel a pressure. She floats as her weight of water would float.

Her motion, her rise and fall, is an oscillation which corresponds to the oscillation of the waves. This vertical motion is the transfer and retransfer of rotative force, her mean level of rotation being still preserved. The wave, rising by force robbed from the passing wind, transfers force to the ship, which also rises; this force expended in the superior level, she pauses a moment, and then falls with accelerated force, till again met by the rising water, — ever floating as the body of water whose place she takes would float.

A picture is beautiful because it bears the impress of the artist, — because on the canvas has been laid the idea of his mind, — because by the drawing, coloring, and combination, he declares the thought which engrossed him as he sat at the easel. Thus nature is beautiful, when in addition to the mere outline, the form, the color, which the eye takes in, the mind perceives also the characters, the language which declares the power and wisdom by which it was constituted.

In this point of view how beautiful is a ship, borne along on the wings of the wind! She is not a mere combination of wood and iron, of cordage and canvas, but she represents and proclaims to us unseen force, — the laws of the elements. Hence the sailing of a ship becomes of absorbing interest; she is full of life, upborne by innate energy, as she bends to the breeze, leaps joyously over the waves, or sternly defies the power of the storm.

We referred, in the first part of this chapter, to the impossibility of explaining what is called specific gravity, according to the theory of the gravitation of fluids. The weight of any mass of matter is supposed to measure exactly the degree of attraction which the earth exercises upon it. Why should the attracting force be lessened when the body is placed in water? If an immersed stone were pressed down by the water, its weight would be increased by the immersion; or, if the fluid pressed upon it equally in every direction, its weight would remain the same as before. If the pressure upon it were upward only, its loss of weight might be accounted for. We explain the loss of weight of a body immersed in water in the following manner: The force which gives certain degree of motion to a body is measurable by the quantity of matter that the moving body contains, each atom receiving its due proportion. It follows that the force required for any determinate degree of velocity depends, not on the volume of the body, but on its density. The force required to move any body, therefore, is in proportion to the quantity of matter which it contains, not to its bulk,—to its atomic structure, not to the space that it occupies. The force needed for the motion of a body in any direction being ascertained, the quantity of matter in the body relative to the space that it occupies can also be ascertained.

The force of rotation in bodies at the same level being in direct proportion to the quantity of matter which they contain, all bodies, large or small, dense or porous, in falling having spare force proportioned to their quantity of matter, should move downward with equal velocity without reference to their bulk in relation to density. This is the case with bodies falling in a vacuum; a feather falls as quickly as a bullet. But in bodies falling through the atmosphere, the greater the bulk relative to density, the less is the comparative velocity, and in falling through water the retardation of velocity is greatly increased. These results are usually thus expressed: "The greater the bulk of the body compared with the matter it contains, the more the air or water resists the fall, and thereby decreases the velocity." Throw away the consideration of velocity which is an effect, and the statement is more simple. In a vacuum, the force of descent relative to the matter is the same in all bodies, large or small, light or heavy; in the atmosphere a part of the force derived from change of level is transferred to the air, and there remains less force for descent; in the water a still greater part is transferred, and still less force remains for descent. The force transferred to the air is exactly that degree which is required to move a volume of air of the bulk of the moving body, through the same extent of space, and with the same velocity as that of the falling body. So the force transferred to the water is that which is requisite to move an equal body of water through the space described by the fall, and with the velocity of the fall. In the descent of a body, therefore, through air or through water, it moves, in opening its path, a volume of the medium through which it passes, of its own bulk, and with its own speed. Its loss of force therefore measures the weight of its own bulk of the fluid through which it passes. The water or air moved may be moved upward, downward, or sideways; but the imparted force is exactly of that intensity, which would have raised a volume of air or water of its own bulk as high as the level from which the falling body descended. Hence the retardation of the fall measures the weight or density of the medium, and conversely the density of a stone is ascertained by being weighed in water. This is an unvarying principle. The seeming discrepancy in the fact that the fall of bodies of great extension of surface as compared with their density is of less velocity, as in the case of the parachute, is because a mass of air must necessarily fall with it, as if the air and parachute constituted one falling body.

If a mass of wood falls in a vacuum, its motion downward is with all the force derived from its change of level. If in air, it will fall, though with less force; for the force derived from its descent is sufficient to give it its own motion, and to raise a quantity of air equal to its own volume. In water, however, it cannot fall, for the force that would be acquired by its descent to a lower level, is not sufficient to move a quantity of water of its own bulk. In the space occupied by the wood, the quantity of matter is less than it would be in the same space if filled with water. The force of the descent of the wood is in proportion to its own density, — to its own quantity of matter. In order to descend, it must open its path. It must transfer enough of its force to move out of the way the surrounding water, — not a quantity at once equal to its own volume, but a quantity proportional to its degree of descent. If the water of equal bulk contains more matter, the spare force of the descent of the wood is not sufficient for any degree of the elevation of water; of course, there can be no descent of the wood. It can only float. Buoyancy depends, therefore, on inferior density compared with the medium in which it is placed, not on the quantity of the medium. If a body placed in water contains more matter than the same volume of water, it falls or sinks. If of equal density with the water, it floats in it at any distance from its surface. The case is the same in air, or in any other medium. By weighing a body first in a vacuum, and afterwards in air, its comparative density can be ascertained. In the floating of a ship, partly in the air, partly in the water, her density is measured by both, that is, her degree of specific weight gives her place between the two elements. Lighter than her bulk of water, heavier than her bulk of air, she neither sinks nor rises, but remains suspended at the junction, at the meeting surfaces, of the ocean and of the atmosphere.

It will be seen that this explanation is somewhat similar to that which is usually given. Those who believe in terrestrial attraction do not think that a stone is less attracted when in water; they attribute its loss of weight to its displacement of water, because the loss is measured by the weight of the quantity of water that would fill the space occupied by the stone. But it is not what a stone excludes, but the quantity of matter which it contains, which determines its weight, and its atoms remain the same whether it is weighed in water or in air. An explanation, therefore, based upon the theory of gravitation, cannot refer to the quantity of water excluded by the stone, for the force with which it is attracted is not lessened thereby. The stone and all that it contains, that is, all that is interior to the medium, is unchanged by the medium. Under the theory of gravitation, there can be no cause for the loss of weight except the action of that which is exterior to the stone; in other words, the degree and kind of the pressure upon it.

But outside there is no conceivable action of the water induced by its gravity, that can possibly result in a loss of weight of the stone equal to the weight of excluded water. Gravitation, we repeat the idea, pressing the water downward upon the stone, will not give the result; gravitation, pressing the water against the stone equally in every direction, will not give it; nor will a current of water moving from above to beneath the stone, for under the law of gravitation it does not require a force equal to the weight of a given quantity of water to move this water when under water. For a reason of the loss of weight, therefore, all, whatever be the theory, are compelled to go from the pressure upon it to the body itself; from the action of external force, to the force pertaining to the stone; from the theory of gravitation, of external attraction, to the force of motion. Our theory, which gives place and position to matter according to its degree of rotative force, does not require to be changed.

There is indeed no reason to be given under the theory of gravitation, why a stone should lose in weight the weight of water which would occupy its place. There is nothing in gravitation by which it can so modify its action on the surrounding water, as to make it press upon the stone in exact proportion to its solid contents. This theory, therefore, cannot explain specific gravity by the attraction of the stone, or by the attraction of the medium; nor by any possible combination of the two. Gravitation cannot attract the stone more or less, or the water more or less, from the fact that the stone and water are in contact. Nor can the result be modified except by mutual pressure, and we have seen that the kind of pressure which gravitation induces is totally inadequate to explain the phenomena.

On the other hand, do not we present an intelligible and distinct explanation? We consider force as having an actual existence. In the case of a body falling through water, we show the source of the power by which it moves, the force of descent from a higher level, and we trace out the division of this force in action. There is so much less in action upon the stone, as is transmitted for the motion of the water necessary for the relative change of place. Specific gravity weighs this force. Weight is the force of motion. It is not an incomprehensible abstraction; it is not an instinct or tendency of matter to approach matter. There is no weight without motion, and the cause of motion is the force present in the moving body. Has not philosophy erred in referring to other matter as the cause of all motion in matter, instead of looking where the motion is for the force which produces it? Why may it not be supposed that matter has power to move itself, as well as that it can induce motion in other matter?

As we have before said, water is elevated by receiving a share of the stone's force of descent; in other words, water goes up because the stone goes down. Often, however, matter descends because other matter ascends. We take a balloon as an instance of this converse action. It is supposed that a balloon goes up from the superior weight of a volume of air of its own size, — that it is crowded or pressed out of its place by a fall of air into its place. Can this be so?

Air either presses downward upon the balloon, or equally upon it in every direction. As we have before said, neither nor both of these pressures could produce elevation. The current of air from above downward is not established until the balloon has commenced its rise; this current is therefore a result of the motion, not its cause. As water is lifted by the fall of a stone into it, so air is moved downward by the ascent of a balloon. It is the peculiarity of construction or of element in the stone which enables it to fall in relation to water; it is the construction or elementary constitution of the hydrogen of the balloon which causes its rise in relation to the atmosphere.

This idea we will illustrate by supposing a balloon held down by a cord, and a dense body of the same size and form held up by a cord, each producing the same tension of the cords which restrain their motion. They are at rest. The air presses on both in the same manner and with the same force. The "tendency" of one to go up, and of the other to go down, comes not from the atmosphere; for its pressure, being on bodies of the same form and size, is of course the same on both. By drawing each of them aside, — by giving them an impulse, — they will swerve to and fro as does a pendulum. The air does not aid this movement, but retards it; for both the swinging bodies lose a part of their force of motion by the "resistance" of the medium in which they oscillate. Does gravity cause this converse oscillation, by depressing one giving it force for elevation, by elevating the other giving it force of depression? When the oscillation is suspended, if the cords which confine them are cut, one will commence its rise, the other its fall, before a particle of air has moved in relation to either. If we suppose two equal receptacles, one filled with hydrogen so that it will rise, the other with a gas heavier than the atmosphere so that it will fall, the external air bearing on each with the same pressure because they are of equal size and similar shape, we cannot avoid the conclusion that the cause of the rise of the one and the fall of the other depends not on the surrounding medium, but on a positive difference in the structure or character of the inclosed gases.

It is impossible to suppose that gravitation attracts the one and repels the other, or that it modifies the pressure of the air around the one to favor its ascent, and around the other to favor its descent. The interchange of force from the medium to the moving body, or from the moving body to the medium, is a result of motion; it is not the impelling cause.

The cause of all motion is the presence of force in the moving body. The degree of rotative force determines the level of rotation. For instance, hydrogen at its formation, at the commencement of its separate existence, possesses an intense molecular force. If unconfined this force changes into progressive motion, and it rises, exhausting its atomic intensity in rotation at the higher level. Its power of elevation, and the height of its rise, depend therefore on the degree of the present force.

But the question returns, what causes some bodies to ascend and others to descend relatively to the medium in which they move? Why does hydrogen change its force of molecular action into progressive motion at the higher level? Why does the stone give up a part of its rotative force and descend? In other words, what is the principle which governs the degree of rotative force, which bodies left free to take their own position will retain?

We have before noticed the fact that cohesion and chemical affinity both depend upon sympathy of molecular action,—on the equality of the diffusion of the present force. It is this which forms all combinations, and determines the separate existence of masses and of volumes. And as the force present in each atom determines its position in relation to other atoms, in their combination forming masses, the same principle determines also the relative position of the masses.

Molecular action is the counterpart of progressive motion; there is a normal ratio between them. The degree of rotative force that any mass or volume can retain if free to change its position, depends upon its capacity for molecular action. This capacity depends upon atomic structure, and atomic structure determines level. The fall or rise of bodies relatively to each other, is determined therefore by the comparative density of the moving masses. One law not only governs combination, but also governs the level of rotation.

Comparative density, — the quantity of matter relatively to the space occupied, — in other words, capacity for the motion of its atoms, determines whether a body rises, or falls, or remains suspended, in any medium. And as all terrestrial bodies move in some medium, or move relatively to other matter, it is the distance between the particles of a body, — its atomic structure, — which causes it to rise or to fall, or which gives it a relatively permanent position. The force of rotation belonging to position is not always the normal force of the structure of the body. If, therefore, there is space for motion, it is moved upward or downward until position conforms the rotative force to the normal degree of molecular force, which is its counterpart. The result is the formation of currents of revolving matter, in which there is an equality of the diffusion of force in relation to *space*, as well as in relation to *density*. The result is, the sympathetic motion of every atom composing the world, — a regular increase of motion from its centre to its outermost circle.

To unfold this principle as far as it appears to us distinctly conceived, would require a greater space than can be given in this outline. It would be out of place also in a treatise designed to present only the very first principles of a theory. We confine ourselves therefore to the thoughts already suggested. A body falls because it is not surrounded by that medium for which its atomic structure fits it; it is not in that stream of revolving matter in which force is normally diffused. It falls when the force gained by descent is not only sufficient for its downward motion, but also sufficient to give the corresponding elevation to its volume of the matter by which it is surrounded. A body rises under analogous conditions, — when it has force not only for its own ascent to a higher level, but enough to move from its path its own volume of the surrounding matter. A body rises or falls according to its present force, — according to its atomic structure or capacity for molecular action.

The equilibrium of force, the normal motion of matter, is preserved by the resolution of the molecular into greater progressive motion, at a superior altitude, or by the decrease of rotative force by the descent of a body. The force pertaining to the atoms and that pertaining to the mass are thus brought into conformity, whatever may be the structure of the mass; the great result is harmony of motion, — the equality of the diffusion of force.

The rise and fall of bodies, therefore, is but an interchange of place between equal volumes of matter. And, while the cause of the motion of a body is within itself, as to the velocity and direction of its motion, it depends upon the medium through which it moves. Matter influences the motion of other matter only by an interchange of force, — by the transmission of motive energy from one body to another. It is force which

> "Sweeps through the dull, dense world, compelling there
> All new successions to the forms they wear."

Thus is the place and position of all things determined, — not by the "appetite of matter" for matter, — not by the desire of one fragment of the world for the presence of another fragment, — not by mutual attraction, — not by a blind instinct of consolidation, kept at times in abeyance by an equally blind instinct of repulsion. It is by the diffusion of force, giving the position of every atom, that the order of the universe is maintained. Change is as normal as permanence, — fluctuation as much a phase of perfection as stability itself. A solitary rain-drop gives us a glimpse of the law, which in its action clothes the earth with verdure; the sudden shooting of a minute crystal into perfect symmetry discloses the power which gives form and cohesion to the world.

This view carries us again to the thought which we have before attempted to elucidate, — to our idea of the unity of force, — to the oneness of that energy, which under many manifestations constitutes the "Life of the Universe." We hear it in the rustling of the forest leaves, in the roar of the ocean, in the thunder of the storm; we see it in the aurora's arch, in the volcano's blaze, in the lightning's flash; we trace it by the oscillations of the barometer, by the tremblings of the magnetic needle, by the action of the voltaic pile; we witness the full majesty of its action in the motions of the spheres.

This thought, which resolves the powers which move all things above and around us into one and the same force, ever circulating, ever inducing perfection, never subsiding into rest, is not a mere poetical idea, a dream of the imagination. The fact of the unity of force, — the identity of its nature under whatever form it may present itself, — is proved by the most accurate experiments, by the severest tests of philosophy. Fill a metallic plate with a crystallizing solution, place a horseshoe magnet under it, crystallization ensues immediately, and the line of crystals formed writes down the direction of the current of magnetic force.

Cohesion, then, is but one phase of force, — symmetry of form one of the results of its action. A rain-drop falls before us. We know from the experiments of Farraday the exact amount of force which formed it from its constituent gases, and by the abstraction of which it will be again resolved into its elements. Further, we trace in its fall the operation of the same force which gives its rotative and revolving motion to the world.

For another illustration, we would recur to the fact that metallic bars, when cooling under certain conditions, are thrown into a state of vibration, and emit a musical note like the sound of an eolian harp, and that the same note is again heard from soft iron when becoming magnetized. Thus the poetry of science is but the revelation of its highest truths. Why should we continue to regard the world's beautiful mystery with but one pervading idea, — that of the power of matter to draw other matter into contact with it? As well might we attempt to solve all the phenomena of animal life by the attraction of the particles of the body for one another, or all the phenomena of thought by bodily sensation, — by atomic wisdom.

CHAPTER VIII.

"HE SUNG THE SECRET SEEDS OF NATURE'S FRAME;
HOW SEAS, AND EARTH, AND AIR, AND ACTIVE FLAME
FELL THROUGH THE MIGHTY VOID, AND IN THEIR FALL
WERE BLINDLY GATHERED IN THIS GOODLY BALL."

Dryden's Virgil.

A FULL belief in the attractive force of matter causes every fact to be seen through this theory as a medium. It therefore gives coloring to facts; as it were, it corrects observation; it makes experiment conform to what it has previously determined ought to be the result; it asks, *how* discrepancies can be explained, not,—Can they be explained?

How different the mental process in regard to a new theory. The question at once changes its form. It is now,—Is it possible that this theory can be true in the very face of this or that disagreement with observed facts? In one case, ingenuity exerts itself in accounting for discrepancies; in the other, in discovering them.

We will refer to one illustration of this idea, though it is somewhat foreign to the immediate subject of the present chapter. It is well known that the pendulum alters its rate over different parts of the surface of the earth, in the same parallels of latitude. In different mines, too, it varies at the same depth. Of course, it might be inferred that the force of gravitation changes in intensity. But, according to theory, gravitation is invariable in relation to the same quantity of matter. Thence it follows, if there be less attraction at any one point, that there are caverns and hollows in the earth; if there be more, the density of the rocks near the surface is far beyond the mean density of the earth; or there are "deflecting causes concealed beneath the surface of the soil." Why the less density of the ocean does not invariably change the rate of the pendulum has not yet been declared. The Shechallien mountain in Scotland deflects a plummet from the vertical direction, and from this swerving the mean density of the earth has been calculated. Other mountains cause little or no perturbation of the attractive force; they are therefore but hollow cones, — though they may have miles of height and miles of base, they are mere shells, — the thin walls of extinct volcanoes! Volcanic islands, on the other hand, generally show "an augmented intensity of the attraction of gravitation;" where there is an exception, it is traced to the influence of some high land on a neighboring coast. Thus a theoretic uniformity is preserved, whatever may be the statement made by facts. The mountain which swerves the plummet is proof of the universal law of the attraction of matter; that which does not offers no disproof, — leads to no doubt of the general principle. It is well that it is so. Theories thus establish order, when without them there would be confusion. It is better to obey a provisional government, though it be not destined to perpetuity, than to be entirely without law. Theories are but steps in progress. How painful the thought that any present knowledge is the highest that can be attained; that any theory is the widest reach of the intellect; that in the future there are in store for us no further interpretations of the wisdom of God as written upon his works!

We return to the examination of the theory of the gravitation of fluids. If a tumbler is filled with water, a piece of paper pressed over its mouth, and the glass then inverted, the water, instead of falling from it, will remain suspended. The slight adhesion of the paper is sufficient to overcome the whole force of the attraction of the earth upon the water. This is explained by the pressure of the air induced by gravitation, which is supposed to act upward against the paper.

But, according to the principles of the gravitation of fluids, this effect could be produced only when there is no air left in the tumbler with the water. It has been stated that, if the smallest quantity of air remained, its pressure above the water would neutralize the pressure of the external air, and the water would fall. But such is not the fact. Let the glass be one half, one quarter, one eighth full of water, and then inverted as before, the water will remain suspended with air over as well as below it. We have often tried this experiment, and called others to witness the fact, that water suspended mid air, with the same pressure above as below, refuses to obey the law of gravitation.

Fill a long tin tube with dry sand, and press on one end a piece of thin paper slightly moistened; let the upper end be open so that the pressure of the air will act both upwards and downwards upon the sand, and thus neutralize itself as in the former case. The sand will not overcome the friction of the sides of the tube and the adhesion of the paper, and will remain suspended, say two pounds of it, against the attraction of gravitation.

Here is a case precisely analogous to that of the water suspended in the tumbler. How is it explained? "Single grains of sand placed on a flat surface do not begin to roll, until the heap is inclined between 30° and 40°, and this is the angle at which pressure is exerted, when this motion is resisted; and the weight of a pyramid, whose sides are inclined at this angle is supported at the base alone, whatever may be the height of the column confined above it," therefore, "the paper supports the sand, though it adheres very slightly;" or, to express the conclusion in other words, the force of the adhesion of the paper to the edges of the tube,—an adhesion, which a very slight force would overcome,—can bear up against the gravitation of several pounds. Why speak of the angle at which sand rolls? It is not necessary for the earth to roll the sand in order to bring it down.

Here are several pounds of sand supported by the friction of the tube, and the adhesion of the damp paper against the action of gravity. This is a plain statement of the fact, and the explanation should be intelligible, without the mystification of the angles of a pile of sand, or of the rolling of sand; for the fact has no reference to rolling, nor to the stability of a heap of sand firmly supported at its base.

The reason for the suspension both of the sand and of the water is the same. It is not the pressure of the atmosphere; for this pressure is neutralized by being above as well as below the suspended matter. The force of a falling body is gained only by its descent. Relatively to the containing vessel, neither the sand nor water has descended. It has acquired no force, and it is not attracted downwards. The glass, raised or depressed by the hand, receives from it the degree of force for its different levels. The water and the glass are as one mass for rotation, the force being equally diffused through them.

The reason that the paper makes the glass and its contents as one mass is obvious. It arrests the beginning of the process of descent. It permits no particle to acquire force of descent. It could not withstand an attraction of gravity; but it prevents any fall by which the contents of the glass can acquire force for the downward motion. Take away the paper, and the sand or water will issue, the surface sinking as stratum by stratum the descending matter makes room for the upper strata to descend. Each stratum for itself gains force of descent by its descent; each stratum quietly waits for its turn. The paper prevents the commencement of this process, and, until it begins, the tube and its contents rotate at any level as one mass, with the rotary force equally diffused through it.

We have thus presented water remaining suspended with a volume of air above it of the same elasticity as that below it, and sand supported in the same manner with the weight of the whole atmosphere pressing upon it above as well as below. We pass from these to other proofs of the want of attractive power in the earth.

If water be pressed upon by a piston, as in a common forcing-pump, with a force of fifty pounds, this pressure will show itself in the strength of the issuing jet. Now take away the piston, and replace it with what under the law of gravitation would be an exact equivalent, fifty pounds of water. The pressure of this water, if it were attracted downward by the earth, would be equal to fifty pounds of external force. But the issuing jet is not increased in nearly the same ratio as when pressure was applied by the piston. The fact is, that the added water has no pressure, and the increase in the force of the jet by its addition, results only from the lengthened column of water, which gives increased force of descent. The increased force of the jet bears an exact mathematical relation to the increased altitude. Before the addition, the velocity of the jet at the orifice was the square root of half the altitude of the water. The addition of the fifty pounds gives a jet, the velocity of which is the square root of half the new altitude. In other words, the force of the jet is the force of rotation unused by the water which has descended in the cylinder, and to get this measure, the mean descent, or half the altitude of the descending column is the element of calculation. The water at rest is not attracted,—does not bear upon the water below it. The addition of the fifty pounds alters the jet, only by increasing the elevation from which the water falls.

But there is another fact relating to the flow of water, which is beautifully illustrative of the views we present. A jet of water, issuing from the orifice of a containing vessel, contracts at a short distance from the orifice. This is technically called the *vena contracta*. The velocity of the issuing jet at this contracted point is the square root of the whole elevation of the water, while, as before stated, the velocity of the jet at the issuing orifice in the side of the vessel is only the square root of half the elevation.

We believe that the increased velocity of issuing water at the *vena contracta* has never been satisfactorily accounted for, and it is admitted that the whole subject of the motion of fluids is still imperfectly understood. We will briefly sketch the history of speculation on this subject, taking our account chiefly from the History of the Inductive Sciences, to which we have so often recurred when we wished for a clear statement of philosophical opinion.

Castelli was the first to assert that the velocity of efflux depended on the depth of the orifice below the surface. Torricelli also stated in 1643, as the result of his experiments, that the fluid would spout nearly to the height of the surface, and inferred that the full velocity is that which a body would have in falling through the depth, and that the velocity is consequently proportional to the depth.

Newton, in his Principia, treated the subject theoretically, that is, according to the theory of gravitation; but "La Grange says that this is the least satisfactory portion of that great work." Newton made experiments different from those of Torricelli; he measured the quantity instead of the velocity of the flow. The velocity inferred from the quantity is only that due to half the depth of the water.

Newton explained the difference by observing the contraction which the jet of water undergoes just after it leaves the orifice. At the orifice the velocity is that due to half the height, at the *vena contracta* it is that due to the whole height. The former velocity regulates the quantity of the discharge; the latter, the force of the jet.

"In the second edition of the Principia, Newton attacked the problem in a manner altogether different from his former one. He there assumes that where a round vessel containing fluid has an orifice at its bottom, the descending fluid may be conceived to be a conoidal mass, which has its base at the surface of the fluid, and its narrow end at the orifice. This portion of the fluid he calls the *cataract*, and supposes that while this part descends, the surrounding parts remain immovable, as if they were frozen; in this way he finds a result agreeing with Torricelli's experiments on the velocity of efflux."

"We must allow that the assumptions by which this result is obtained," continues Whewell, "are somewhat arbitrary; and those which Newton introduces in attempting to connect the problem of issuing fluids with that of the resistance of a body moving in a fluid are no less so. Even up to this time mathematicians have not been able to reduce problems concerning the motion of fluids to exact calculation, without introducing some steps of this arbitrary kind. . . . . Hence the science of the motion of fluids, unlike all the other primary departments of mechanics, is a subject on which we still need experiments to point out the fundamental principles."

We would ask the question, — may it not be found that the theories connected with this subject require an examination? The facts will not change their character; the jet of water will ever continue to bear the same relation to the depth of the column as in the days of Torricelli and Newton. The quantity of water flowing out, as also its velocity at the orifice, will be that due to a fall from one half the height of the column; its velocity at the *vena contracta*, and the height of its rise, will retain their relation to the whole depth of the column, — that is, at the orifice the velocity will ever be the square root of the half altitude, at the *vena contracta*, the square root of the entire altitude.

If water in the containing vessel were the subject of terrestrial attraction, the force of the jet at the issuing orifice would be determined by the whole weight of the water. As we have seen, this is not the case. Or if the attraction acted only on a column of the same area as the opening, the force of the jet would be determined by the whole height of the column. But, as we have before said, it is found on experiment, that the force of the jet at the orifice and the quantity of water issuing, is that which is due to one half the height only. On our principles, the force of the jet, being the force freed by descent, could be related only to one half the height; for the water at the orifice would have no force of descent within the vessel, while the water at the surface would have the force due to the whole distance from the top to the aperture. Of course, the force of the jet would be the mean force of the column, or the force of the descent from one half the height. This appears almost perfect proof that it is force from descent, and not an attractive force, which gives the velocity of the jet, and determines the quantity of issuing water. If gravitation brought down the water, the whole of the column being equally attracted, the force of the jet would be that due to the whole height.

As the force of the jet, therefore, is determined, not by the quantity, but by the height of the water in the containing vessel, it is force of descent only which governs the intensity of the flow. There can be no difference of opinion thus far. Action on the jet is confined to the water which is in motion. In fact, the impelling power is usually stated to be the force of descent induced by the action of gravitation.

But this force of descent does not account for the increased velocity at the *vena contracta.* It is in its full intensity at the moment the water issues from the orifice. All the power produced by the change of level has accumulated at the aperture, and after the water has issued it can receive no farther increment of force from descent within. Its increase of velocity after it has left the vessel, is not therefore from the force of descent. Nor is the motion accelerated *because* the jet is contracted; as the quantity of water remains the same, the size of the stream depends on the velocity of its passage. The contraction is occasioned by the more rapid flow of the particles of water; it does not cause the rapidity. The jet occupies less space because it is accelerated; if it were retarded, it would increase in circumference. Most unquestionably, therefore, the water after leaving the vessel, after its possession of the full impulse from the fall within, is acted upon by a new impelling power. The jet is doubled in velocity, and by a cause which commences its action after the issue of the water. The increase of velocity at the *vena contracta* cannot be caused by the force of descent, whether this force of descent comes from gravitation, or from change of the level of rotation.

But the increase is measured by the depth of the water from which the jet proceeds. There is then a difference in the state or condition of water according to its depth. This difference is in the degree of molecular action. As we have before observed, we believe that the force of rotation decreases from the surface of a fluid downward, and that molecular force increases proportionally in the same direction, — that the one is the complement of the other, giving to the whole volume of the fluid an equal diffusion of force. The proofs of this theory seem to multiply as we proceed. And the increase of the velocity of the jet at the *vena contracta*, not only shows the existence of this molecular force, but gives us the measure of its intensity, and its intensity thus determined presents it as the counterpart of the force gained by descent.

At the orifice we have a force of flow which is due only to the mean descent of the column of water; at the *vena contracta* we have an additional force exactly equivalent to it, so that at this point the force is equal to that which would be disengaged by descent from the whole height of the column. The two forces are therefore the exact counterparts of each other. If the stratum of water nearest the orifice has no force of descent, it is compensated by a greater molecular force; if the stratum at the upper surface has less molecular energy, it is compensated by its greater rotary force. The velocity of the jet at the orifice is the result of only one force; at the *vena contracta* it is that of the sum of the two forces. At the orifice there is only the force of descent; at the *vena contracta* the force of the molecular action is added to it. The change of level throws the force of descent into action on the jet; the escape from the depth changes the molecular force of that depth into progressive motion at a short distance from the orifice.

It is also found that more water discharges itself from a reservoir through a short pipe, than through a naked aperture of the same diameter as the pipe. This fact has been explained by the supposition, that "the issuing particles coming from all sides to escape, cross and impede each other in rushing through a simple opening, as is proved by the narrow neck which the jet exhibits a little beyond the opening; but in a tube this narrowing of the jet cannot happen without leaving a vacuum around the part, and the pressure of the atmosphere resisting the vacuum causes a quicker flow." Surely the results of atmospheric pressure induced by the attraction of gravitation are many and wonderful! Here we have an increased efflux of water by the pressure of air against it!

The contraction, as we have observed, is caused by the increased velocity of the jet; for of course from the more rapid motion of the water, there is less of it at any one moment at any one point of its path. The effect of the short tube is to prevent the full contraction, by the adhesion of the water to its interior surface; but, as this adhesion does not overcome all the gain of velocity by the transfer of molecular into progressive motion, the force, instead of acting at a distance from the aperture, diffuses itself and makes an average velocity of all the water in the pipe. By friction the additional velocity is thrown back to the aperture, increasing the speed of the fluid through it. To obtain the greatest possible increase of flow, the tube should first narrow to the point where the *vena contracta* forms in the natural jet; this gives time for the transfer of atomic into progressive motion; then it should widen from this point so that the water may spread itself with the increased flow. It will be seen that the effect of this appliance, — technically the conical ajutage, — is to make the velocity of the *vena contracta* the same with that at the point of efflux, — the velocity at the opening. Increased velocity at the aperture emits a greater quantity at any one time than would pass with a less velocity. The greater the efflux, the greater the force of descent, and the greater the addition to the flow by the action of the molecular force.

It appears to us a clear explanation of the effect of conical ajutages, to consider them as the means by which the force that without them would act only away from the opening, increasing the velocity at a distance from it, is made to act at the opening where the greater velocity passes through the aperture a greater quantity of water in a given time. By the retardation of the flow of the water by the friction of a pipe, the force of the flow is not annihilated. It is still in being, and if its onward action be impeded, it is thrown back toward the opening, giving passage to more water from the reservoir. And with more water comes its own force of propulsion, and of reflex influence. So the gain in the flow is limited only by the quantity of molecular action which the depth of the reservoir can supply, and that is limited to what would be the force of descent if all the issuing water fell from the upper surface of the reservoir.

By the theory of gravitation the efflux of water from the

opening in the reservoir is represented as the effect of terrestrial attraction, and yet there is an admitted impossibility of giving by this theory any thing which approximates to a clear explanation. Why, if the water in the reservoir is attracted, and the result of this attraction is an equal pressure in every direction, does not the jet respond to this pressure when it answers at once to the presence of any external force so applied as to bear upon the particles of the water in any direction? If there can be clear proof of any position whatever, there is this clear proof that gravitation does not in the least degree press together the particles of a fluid.

Why again is the issuing jet imputed to the force of descent induced by gravitation, when it is so palpably evident that a part of the velocity of the jet comes after the descent has exerted its full influence, and when the quantity of water discharged is increased by an appliance which cannot possibly be supposed to have the least effect in increasing the force of descent? And why impute the whole force of the jet to a cause, which if it act at all, can produce only a moiety of this force?

On the other hand, have not we a theory which gives a distinct idea of the cause of these phenomena? One moiety of the force of the discharge comes from the change of level, — the unused rotary force resolved into other motion; the other moiety, from the resolution of molecular force into progressive motion. United they are sufficient again to raise the water, as it flows into another receptacle, to the height from which it fell, — the spare force of rotation being again transferred to rotation, the molecular force being again translated into the atomic action due to a depth equal to that from which the water issued.

Practical men recognize these principles in the construction of their machinery. The molecular force distributed as the counterpart of the rotary, increases regularly from the lower surface upward. As a cone, with its base at the lower stratum, and its apex at the upper, must be divided at one third of its altitude, to leave equal quantities in each portion, so the molecular force is divided; the centre or equilibrium of power in every column, whatever its height, is at one third of the altitude from the lower surface. To support a sluice, or flood-gate, or embankment, the support is made to press at one third of the height.

A practical millwright, on being asked if it made any difference how the water was let on to his wheel, replied, "On the overshot or vertical wheel it does not make any difference; in whatever way the water runs out, I get its whole weight as it strikes the wheel; but I confine that which issues from below on the central, or horizontal wheel, so that it shall not spread until it strikes the wheel, for thereby I get the 'spring' of the water." "But you know that water is incompressible; how can it have a spring?" "It may be so," replied the mechanic, "but I know that water issuing from below has a spring to it."

We are far from having exhausted this branch of the subject; but, as it has already occupied an undue share perhaps of our space, we pass to a cursory consideration of the ebullition of water, and of the action of steam.

By recognizing the action of molecular force, we are enabled to comprehend a peculiarity of water in its conduction of heat. It is almost impossible to heat water from the surface downwards. Heat can hardly be propagated in this direction through the fluid. If it is applied at the surface, the upper stratum only receives the force, and is converted by it into vapor, which passes off. But when the lower stratum receives the force, it distributes it normally through the volume, the molecular action being gradually increased until ebullition, that is, vaporization from every stratum, ensues. The lower strata, not the surface, regulate molecular action; without the application of artificial heat, it is additional depth which increases its intensity. The force thus acting in the lower stratum determines that of all the succeeding strata; therefore, it is only by artificially increasing the force of the lower surface, that this increase can be extended through the whole volume. Our views receive confirmation from the fact, that water artificially heated is always hottest at the bottom of the vessel; in very deep boilers, the liquid is much more heated than it can be in smaller vessels.

When there is no application of artificial heat, the molecular force present in water depends upon the degree of rotary force. Therefore water boils with less added heat, according to the degree of its elevation. At a great altitude, as water is already in possession of a high degree of rotary force, it does not require so much additional force to bring the molecular action to that degree of intensity which changes it into steam. At the summit of Mont Blanc water boils at 180° of Fahrenheit, — 32° less than at the level of the ocean. By the changes of rotary force upon the surface of the earth, vaporization from water, and even from snow and ice, is induced and suspended. This vapor is not expansive, having space for its range of molecular motion.

Steam has, of course, a most intense molecular action, and, when generated under confinement, it has not sufficient range of space for its normal motion. Hence its expansive power.

Ebullition does not, in the least degree, depend upon atmospheric pressure. Air, in its normal condition, does not act upon water. But if air be artificially condensed over water, as in a steam-boiler for, instance, by the rising of steam in it, there is then a pressure of the air upon the water, and there is less space for the water to diffuse itself. When closely confined, without space for enlargement, water can be heated red hot without passing into vapor. The degree of the confinement therefore measures the resistance to the change of water into steam. In practice, the resistance is the impelling force which moves the piston. Hence the importance of confining steam in as small a compass as possible, and not allowing it to expand except in that direction in which, by the act of expansion, it can move the machinery. The force of steam is its demand for greater space for its molecular action. That attained, it is no longer the source of motive power. It cannot be recompressed without the use of as much force as it would give out by its reenlargement. Great expense is often incurred by the want of a clear understanding of this principle. In an artificial vacuum on the other hand, there is greater space for the molecular action of the remnant of air than is normal to it, and if water is placed in the vacuum, the equilibrium is restored by its vaporization. So great at times is this vaporization, that there is not force enough left in the remaining water to preserve its fluidity, and it becomes frozen.

The increased capacity of water for heat in proportion to its depth, to which we have alluded, is of course attributed to the pressure of the upper water, which is supposed to prevent the escape by vaporization of the heat from the pressed strata. Can this be a well-founded opinion, since gravitation neither directly nor by pressure can have power over heat?

Nor is it necessary that there should be vaporization to produce an equal diffusion of heat throughout any one volume of fluid. The lower strata retain the augmented heat on the same principle on which they retain the molecular force. The additional force of heat is diffused in a decreasing ratio from the bottom to the surface of the fluid.

We will advert to another of the supposed results of the gravitating power. It is believed that, if the pressure of the atmosphere were withdrawn, many substances now solid would become fluid, and many fluids, aeriform; that water with a slight increase of heat, would change itself into vapor, were it not for the weight of the air upon it; and that in fact it is by the gravitation of the atmosphere that even the ocean remains in its place.

There is one striking peculiarity in the philosophy of gravitation. According to it no matter in the universe moves or keeps in place without the control of other matter. Every particle is endowed with energy to give motion to other particles; but it has none for its own motion. Thus nothing is able to take its own position, to be what and where it is, from itself, from its inherent fitness; but every mass is perfectly competent to press into place adjacent masses or volumes, and these in their turn to press it into proper position. And this is deemed true, not of adjacent masses only, but of masses millions of miles distant, and that against the action of the force with which they were endowed at their creation. Thus one sphere determines the orbit of another, itself being guided in its turn by an emanation from some other centre of influence, and hence there is order and permanence in the system of worlds.

It may be said that this feature of the theory of universal gravitation is not objectionable, because the order of nature may be as firmly preserved by the indirect action of matter, as by the immediate power of every atom to move in its destined path, and to preserve its own proper position.

It is true that, whatever be the character of our philosophy, it cannot change the perfect and the permanent. The worlds will roll on in their orbits and fulfil their appointed rounds, whether we believe them endowed directly with power from the hands of their Maker, or think of them as under the control of other parts of the creation. But theories have value in proportion to their effect in strengthening our conceptions of the wisdom and power by which the universe is continually upheld, — in leading our minds upward to the Great First Cause, instead of allowing them to rest in the finite and the created. The first work of philosophy is to divide the incomprehensible from that which can be clearly apprehended, that the one may be made the subject of investigation, the other be used only to lift our thoughts to the wisdom and power of God. Would that every theory were destroyed which furnished to the mind even the appearance of a resting place between the Creator and his works.

By referring the motions and positions of matter to mutual action, we receive no clear idea, we take no step in advance. We still rest on the power of matter to induce motion, — to do that indirectly which we are not willing to admit that it can do directly. To us, and perhaps to other minds, the theory that every part of the universe, each atom, volume, mass, is guided by force present with it, unaided by the influence or mediation of other matter, has a value because it strengthens our conviction of the perfectness of the creation, — because it leads us to recognize in all the changes of matter the Guiding Hand which sustains, controls, and directs it. It presents to us every element as in its place by reason of its construction, — as discharging its functions by the inherent fitness of its constitution, — by a fitness not borrowed from other matter, but imparted to it at once from the Source of all perfection. It brings before us general order from the perfectness of each part. We recognize in harmony of action, as it were a sympathetic movement, not a compelled conformity, — not the equilibrium of well balanced opposition. We think that every part of the structure of the worlds stands firm because it was so created, — not because it is supported by the insecure balance of opposite leanings from the true direction.

The ocean appears to us as a most beautiful instance of an element, perfect in its adaptation to its place, — in its fitness to fulfil all the purposes of its creation. It is not held to its bed by the pressure of the atmosphere upon it, but works as if in sympathy with it; it is not chained down to its depths, but, as if exultingly, it lavishes its force to aid in the harmony of the world's movement. It seems to rejoice that it is, and that it acts.* Though by its constitution so

* Since writing the above we have met with the following beautiful passage, in an article from Rev. A. P. Peabody, in a recent number of the North American Review, which has given us so much pleasure that we cannot forbear transcribing it: "We are reminded in this connection of an apostrophe to the sea, in the Prometheus of Æschylus, which, in four words conveys much more to our apprehension, and reflects far more fully our own unutterable emotion in the frequent survey of ocean scenery, than could be done by volumes of the most glowing, eloquent, passionate description. Whenever we look upon the ocean, whether from deck, beach, or crag, by sunshine, moonlight, or its own phosphorescent glow, we find ourselves absolutely haunted, sprighted, by these words, as they pulse upon the inward ear in unison with the rhythm of the waves. The passage to which we refer is that where Prometheus, in calling on all nature to witness his cruel wrongs at the hand of Jupiter, addresses the sea as *ποντίων*

nearly poised between the solid and the aeriform state, it is in no peril, from the vicissitudes of temperature, of losing its form and condition by passing into either extreme. It is not in dangerous equilibrium. Its own construction protects it on either hand, and its inherent safeguard strengthens as the volume of water deepens. A mere surface stratum only can harden into ice, or pass away in vapor. The heat of the tropical sun cannot penetrate it; under its direct rays are thousands of miles of ocean surface with a permanent and uniform temperature, and water drawn from the depths, even under the equator, has the icy coldness of the water of the polar seas. The unused force of rotation and revolution, laid up as in a storehouse of strength, is securely held in the molecular action, so that by degrees only it ascends to the surface and passes off in vapor; or by conduction to the air as heat mellows the rigor of winter on the sea-girt shore. So securely is this force held, that from the equator to 48° north and south of it, the mean temperature is but a few degrees above that of the air. Yet we know that it has force proportional to its depth; for the thermometer will determine the presence of shoal places as well as the sounding lead; and the cold water of shallows condenses the vapor of the air over them, so that often, clouds, sharp and well defined, map out accurately the form

κυμάτων ἀνήριθμον γέλασμα. This cannot be transfused without damage into another language.

> ''Tis odor fled,
> As soon as shed.'

We know not how to convey in current English the multiform unity indicated by the original. '*The innumerable laugh of the sea-waves,*' is literal, but awkward. '*The many twinkling smile of ocean,*' (which we copy from a Lexicon,) is preferable on the score of euphony, but less adequate to the sense."

of the banks beneath. The force regularly set free from the ocean preserves the equilibrium of the atmosphere over it, so that, far away from land, there are not the frequent and sudden irregularities, the disturbances from conflicting winds, which occur where islands break up the continuity and interrupt the regular transmission of force.

We referred in a former chapter to the ocean as being a storehouse of force, — as the conducting element by which the unused energies of rotation and revolution are redistributed. How perfectly adapted for this purpose are the oceans from their position! There is no unbroken circuit of land passing round the earth as a barrier against the passage of force from one hemisphere to the other. There are few broad surfaces of land which are not broken up by intruding bays or inland seas. North and south around the poles, the Arctic and Antarctic Oceans open free communion; for there are changes about the magnetic poles demanding transfers of force for the restoration of an equilibrium. And from the equator to the poles on either hand, between which the greatest difference of temperature and of rotative force exists, by the broad sheets of the Atlantic and Pacific stretching from pole to pole, is the communication kept fully open.

We now pass to the consideration of the atmosphere. If a weight bearing upon the surface of the ocean were necessary to prevent it from passing off by vaporization, the atmosphere, as it is constituted, could not supply this pressure. As the air exists in its normal state, it is without power to bear with force against an adjoining volume of other matter. Air can press by its elastic power alone, and it manifests elasticity only when there is not sufficient space for its molecular action. Its elasticity becomes apparent by compression, — it is the throwing back the force which disturbs the normal position of its particles. Like steam, air must have its molecular action artificially repressed by confinement in order to manifest its elastic force. The only pressure of aeriform bodies is from want of space for the action of the force present with the atoms.

As the atmosphere exists in its natural state, its molecular force is determined, as in the case of the denser fluid, water, by its degree of rotary force. Its atomic force therefore increases in every stratum, from the highest to the lowest, as the counterpart of its progressive force. We cannot see a current of heated air pass upward without noticing in its vibrations the intensity of its molecular action, which, as it ascends, is absorbed in rotary motion at a higher level. This atomic force, constituting elasticity when under confinement, when unconfined has its normal space, — has all the room that it needs for its motion. Of course, it cannot bear upon or press any matter in contact with it. We might as well look for motive power by the expansion of the free vapor rising from the surface of the ocean, as to seek in unconfined air for the power of pressure. In both cases the particles have freedom and space for motion without impingement upon each other, or upon surrounding matter. Nature does all things well. There is no tendency in matter to escape from her adjustments. There is no crowding, — no contest for space. All things conform one to another. Force, matter, space, are in due relation. There may be an occasional oscillation, — for a time a limited swerving from an accustomed relation, but how soon the storm subsides! It is not a struggle for mastery, the continued battle of contending forces, which gives the peace of nature. It is, as we have said, the constituted fitness of all things for their assigned place and office.

Air and water rest in contact without mutual pressure because they are in place. If one causes motion in the other, it is by the transfer of force. When air is artificially condensed the force of elasticity manifests itself, and then there is pressure; but, as this is the effect of the unnatural condition, it does not outlast the confinement. Set free, the condensed air rushes instantly into the more rarefied, diffuses itself where there is relatively greater space for motion, and an equilibrium is at once restored. What is true of a smaller, isolated volume of air, is true of the whole atmosphere. The law of nature is uniform; the same force which diffuses a confined portion when it is set free, is ever in unobserved action, so that the whole element is in equilibrium in relation to its own particles, and in its relation to other matter; the law of its nature forbids a permanent and continuous pressure upon other matter.

That air is elastic in proportion to its density has been proved by the most accurate experiments, performed by a commission of the French Academy of Science. To state the fact in other words, the force with which air will diffuse itself is in proportion to its density compared with that of the surrounding air. Elasticity, therefore, is the power of diffusion, by which the pressure ceases to be, — by which it is annihilated in consequence of the return of the whole volume to its normal condition. Thus, the elasticity of a bent bow, so soon as it is straightened, no longer manifests its presence, for the atoms of the bow have returned to their normal state and are at rest.

What view of this subject does the theory of gravitation

present? It supposes a continued, a permanent pressure of the particles of air, stratum on stratum, causing differing degrees of density in every layer, the pressure decreasing from the surface of the earth upward till the air gradually fades away into the vacancy of space. It gives to air a continued struggle for space,—an abiding elasticity,—an elasticity acting in every direction, bearing upward as well as downward. It presents the atmosphere as a series of springs, increasing in the power of reaction from above downward, and all of these bent springs secured at one end, only with nothing to resist the expansion of the other end,—the elasticity of the air increasing with its density, its density increasing with its depth, and this elasticity acting with equal force in every direction! We know that the expansive power of the air throws the more dense portions upward into the more rarefied. What therefore keeps down the atmosphere, which, according to theory, presses upward as well as downward, against itself and all other matter, with a force of fifteen pounds to the inch?

We cannot avoid the belief that the views which we have presented accord more nearly with the facts. The air is neither drawn down to the earth by attraction, nor repelled from it by its own elasticity. It holds its place, because it is so constituted as to retain the due force of rotation for its level of rotation. It has no tendency to escape into the regions of space, nor to crowd into the bed of the ocean. It occupies its allotted place without an effort to expand or to contract its volume. It is without elasticity when uncompressed,—without weight when resting at one level. Its motion is from the reception of force,—its permanency, because God gave to it the atomic and elementary construction which fits it for its position, and for its office in the economy of nature. It did not fit itself for its place and function, nor is it guarded and restrained by other matter as dead and senseless as itself.

That gravitation cannot by its force draw down the atmosphere, so that its density shall increase, stratum after stratum, from above downward, or if the atmosphere exists with this increasing ratio of density, that it is not the action of gravitation which retains it in its position, is distinctly proved by the operation of the air-pump; for the comparatively denser air of the receiver invariably rises, against the attraction of the earth, into the rarer medium between the valves. Thus against gravitation would the denser air near the earth rise up into the more rarefied strata above, till an equilibrium of density was established. "We know not," says Dr. Arnott, "at what distance the gravity of the particles of air becomes just a balance to their repulsive force, and therefore know not what the degree of rarity is at the top of our atmosphere, but we see that it must be exceedingly great from the fact that the air left in the receiver of an air-pump has still spring enough to lift the valve of the pump, when there is less than one thousandth of the original quantity remaining." If the air in the receiver has "spring" enough left for its elevation, though so vastly reduced in density, why is not the spring of the denser strata of the atmosphere a sufficient force for their elevation into the more rarefied?

This illustration introduces us to the next branch of our subject, — the vacuum. We do not entertain the idea that nature abhors a void place, but we believe, to speak metaphorically, that nature abhors and resists the infraction of her laws, one of which is, that force shall be equably diffused according to altitude. Force, in distributing itself, will carry into a void place the adjacent matter, so that in the onward current of rotation there shall be no vacant space. The fall or rise of matter from the stream at one level into the stream at another, is but the interchange of place of equal volumes. In the creation of a vacuum, however, there is a change without the equipoising change, and the filling of a vacuum, whenever it takes place, is the same act as if the equipoise had been preserved when the matter was first removed. Thus vacua are filled, let the time be never so remote from their formation, whenever the adjacent matter is susceptible of the change of form necessary for the purpose, and filled with the same degree of force with which they were made void. Air, extensible without limit, will always fill a vacuum, as will water and other fluids to the limit of their extension; solids will be forced against or into it to the degree that their form and elasticity will permit.

To repeat the idea; the force used to create a vacuum measures the force with which it is filled. For instance, if the air be drawn from a vessel, the force which will cause other air to rush into it, is equal in intensity to that which displaced the air. The force required to fill a vacuum is not determined by the gravity of the external air, for that must be invariable. Matter lifted up does not increase in weight according to the distance to which it is lifted, nor has it more weight of air over it when elevated one foot than when raised six inches; yet the higher a piston is lifted in making a vacuum, the more force is required, and with the more force does it return. Suppose a tube in which, by the elevation of a piston, a theoretically perfect vacuum is formed. The air is believed to press upon every inch of the surface of the tube and piston with a force of fifteen pounds. Raise the piston one foot, the air bears on its outside surface fifteen pounds per inch; — raise it six feet, the weight of the atmosphere over it has not been increased; its pressure remains the same, but the higher it is drawn the more force has been required to lift it, and the greater is the force with which it returns. The strength of the "fuga vacui" is not therefore caused by the unchanging weight of the atmosphere. It is measured by the force which produced the void place, in other words, by the extent of the vacuum, and of course does not depend on the unvarying weight of the column of air over it.

To vary the illustration we would refer to the "Magdeburgh hemispheres," as they are called. Two hollow half globes of metal are fitted to each other, so that their lips when touching may be air-tight. Exhaust the air within by an air-pump, and then "a force is required to separate them of as many times fifteen pounds as there are square inches *in the area of the mouth*, or at the surface of the division of the upper and lower hemispheres." The force holding them together is not therefore the gravity of the atmosphere; for, if the vacuum were perfect, the external pressure would be in proportion to the surface exposed, — to the surface area of the sphere, not, as is the case, to the size of the vacuum measured by its diameter. It is said that, when these hemispheres were first exhibited, the inventor had a pair made of a foot in diameter, and that six horses were unable to pull them asunder. Suppose the metal to be thicker, so that the vacuum would be only half the size. We know that much less force would be required to pull them apart, and yet the weight of the atmosphere around them would remain unchanged.

Thrust down a piston so as to condense the confined air beneath it, it will be driven back with the force with which it was thrust down. Elevate a piston so as to rarefy the confined air, and it will be drawn down with the force with which it was elevated. The force of the readjustment is related only to the void place, not to the weight of surrounding air,—to the extent of the disarrangement, by no means to the weight of the medium in which the void place is made. There is one degree of force pervading the atmosphere, giving to each of its atoms its relative position. Let there be an abnormal position of the particles of any volume of air, and the force that gave the normal position of the atoms, will be the measure of the force required to restore it to its former state. In other words, the force of position is invariable. But let it be granted that it is by the pressure of the air that a vacuum is filled,—the pressure can be caused only by the elasticity of the air, or by the power by which it enlarges its volume wherever there is space. If we allow the air to have this elasticity, every fact of its action relative to the vacuum is explained, and the elasticity could exist as well without as with the attraction of the earth. Gravitation therefore is not the essential condition for the pressing of the atmosphere into a void place. Elasticity in fact supplies a better explanation, and this may exist without attraction.

But we consider the atmosphere in its normal state as without elasticity. The action of elasticity being only induced by a disturbance of the relative density of its parts, its action, so far from being permanent, is remedial, curative, leading to equal diffusion of the atoms of the atmosphere through space. Therefore the air is of uniform density,—therefore are the void places filled. Imagine in the ocean an unfilled spot, surrounded by water moving in rotation with a force of one thousand miles an hour, the particles of water not pressing on one another, the force being equally diffused, each atom having its exact proportion; — into this void place the water will enter, and it will be filled, force diffusing itself equally in the whole stream, and at the same time equally diffusing the atoms of the stream. Thus it requires great strength in the walls of the vacuum to keep out the flowing stream of air; the force by which nature seeks to fill it is the force by which the matter of the universe is kept in constant motion.

This view impresses the mind with the permanency of nature, with the never-varying intensity of her force. However long a vacuum may continue, the force of the return will be the force of withdrawal, and thus, while the position and movement of the sphere is ever rigidly determined, so is the position and movement of the most minute fragment, of the very atoms composing the fragment, by an ever acting, unchanging power. Minor changes take place, — there is a limited elasticity which permits condensation and rarefaction, alterations of form, variation of motion; but the law preserving the general harmony is never repealed, — it remains in constant operation. The force that disarranges measures the force which readjusts; the void place is filled to restore perfection, completeness, but never filled by the disturbance of other elements. They too have their laws, and a void in the air, can be filled with water to a limited extent only; for the energy of nature is used for preservation alone.

In passing, we turn our attention for a moment to the level of water when in contact with the atmosphere. The even surface of an undisturbed volume of water, — the well defined line which gives distinctness to the volumes of the separate fluids, — has been usually traced to the action of gravitation.

In speaking of chemical affinity, we referred to the equal

diffusion of force among combining atoms as the necessary condition of combination. We look upon matter as composed of simple elements, and upon these elements as possessing different relative capacities for the retention of force. If the simple elements in contact do not equally diffuse the present force, there cannot be the harmonious molecular action which alone can unite them as one mass or volume; if the force be equally diffused, there ensues that sympathy of atomic action which is the bond of cohesion. Without this provision contact would determine combination, and in the place of established order, of distinctly characterized masses, all things would be mingled in chaotic confusion. There would be

> "Neither sea, nor shore, nor air, nor fire,
> But all these in their pregnant causes mixed
> Confusedly."

While we believe that there is a certain intensity of force and its equal diffusion, which locks together the fitting elements of every structure, — which produces the various and distinctly characterized solids and fluids; so after their construction, by the operation of the same principle, are they kept separate and distinct. Air and water, for instance, though they may be somewhat mingled mechanically, though for a time their meeting surfaces may be to a degree made irregular, continue to be distinct volumes, ever returning to a fixed relative position. When at rest their line of separation is a perfect level; the position of one is invariably above the other. It is because the one in all the transfers of force retains permanently the greater share, that its level of rotation is superior.

If a vessel containing water has pipes issuing from it of various shapes and diameters, the volume of water will retain the same level in all of its divisions. The water receives that degree of force only, which, while in contact with air, gives it its place below the air, and this force being equally diffused throughout the volume, the force of rotation and the force of molecular action are the same in every part, resulting in a uniform surface level. It is from this principle, that issuing streams may rise to the altitude of their fountain,—that water running through conducting pipes, whatever may be their enlargement or contraction, their length or windings, knows the level belonging to its volume. We cannot conceive of this uniform level being produced by the downward tendency of the particles of the water. But the reason of it appears evident when we consider it as the outer circle of rotation, determinately fixed by the degree of its impelling force. It is from this principle that the vast ocean, constituting almost a liquid world, though agitated by winds, though swollen by tides, though traversed by wide-extending currents, preserves the curve of its surface,—the alternations marked upon its shores showing that its changes, like all changes in nature, are limited and periodic. The ocean is not a collection of particles gravitating independently of each other, nor is the shape of its volume preserved by the tendency of each atom to the centre of attraction. It is a whole, animated as it were with one life diffused through every part.

An explanation of the water level by the theory of gravitation has been attempted, and the explanation has been accepted, not as sufficient and satisfactory, but as the best possible in accordance with the theory. Suppose a vessel containing one thousand pounds of water, with a pipe extending upwards from its lower part of a size sufficient to contain only five pounds; the water stands at the same level in both.

One would suppose that the weight of water in the larger vessel,—its constant tendency to descend,—would force the small quantity upward so that it would escape from the open pipe. Gravitation asserts that the small column is acted upon only by a column in the large vessel of equal area to itself, and that all the water except this column is held in place by resting on the sides and bottom of the containing vessel. But, as we have seen, any downward force actually applied to the water is not resisted by the vessel, but shows itself at once by the issue of water from the orifice. We believe that the same level is maintained by both columns of water, because there is no downward tendency in either,—because the water is one volume, and has its rotative force equally diffused through all its parts.

We pass to some further considerations respecting atmospheric pressure. The theory of gravitation assigns to the atmosphere, as we have several times had occasion to repeat, a pressure of fifteen pounds on every inch of surface with which it is in contact. It therefore bears upon the body of a man of average size with a force of thirty thousand pounds. If this pressure were not purely theoretical, it would be felt. We should be conscious of it; we should not be compelled to go to books for our first information respecting it. Its intensity would be too great for concealment; it would declare its presence by its effects. That it is not felt is accounted for by the assertion that it is equal on every part of the body. "There is," says a popular writer on philosophy, "a difficulty at first in believing that a man's body bears this pressure while he remains altogether insensible to it. But such is the fact, and the reason of his not feeling it is that it is perfectly uniform all around. Fishes are at case in a greater weight of water, and men walk on the earth without discovering a heavy atmosphere about them, which, however, will instantly crush together the sides of a square glass bottle emptied by the air-pump, or even of a thick iron boiler, left for a moment by any accident without the counteracting internal support of steam or air. . . . . In general, the pressure on one side of a body is balanced by the equal pressure on the other, and it is on this account that philosophers were so long in discovering it at all, and that half-informed persons are still disposed to doubt its existence."

We must number ourselves with those who still doubt. How its uniformity, its equal distribution on every part of the surface of the body, exterior and interior, can destroy this enormous pressure, or rather how it operates to prevent its being sensible, has never been explained. The equality of its action could not annihilate the force. If the air presses on all sides of the body, it must press together the sensitive particles constituting the body.

But it is well known that the human frame is so constituted that it is impossible for the air to exert its pressure uniformly. There are cavities in the body which contain air. Now as the pressure on the outside varies in intensity, sometimes a thousand pounds in an hour, according to the indications of the barometer, what is to enable the small quantities inclosed to maintain so nearly a counteracting pressure, that in the continual variations of weight outside no sensation is perceived? In the cavity of the ear, for instance, there is air which is supposed to resist the outward weight; but in order to resist this changing burden, to maintain the equilibrium which prevents it from being felt, this internal air must continually change its own pressure.

To do this it must be increased or diminished, and the flowing of air into and out of the orifice would constitute it a perfect barometer.

But while some organs of the body have cavities containing air, which may assist them in supporting the external pressure, there are others, the brain for instance, which have to bear all the changes of pressure without any internal support. A well educated physician remarks, "How the delicate organs of the human body bear the pressure which gravitation assigns I know not. It is not because of contained air which resists the pressure. Air within the organs would be death. A very slight pressure on the brain,—that of a few ounces only,—is attended with loss of consciousness, yet the varying density of the air changes the pressure upon it by many pounds, and I am yet to learn that the brain contains air, which, by corresponding changes, could neutralize the exterior pressure."

But in their eagerness to prove the existence of atmospheric pressure, some philosophers, forgetting that the principle of its equable pressure forbids any perceptible result from it, either for good or evil, assert that it exerts an important influence on the animal organization. Doctor Arnott, whom we have before quoted, remarks, "The atmospheric pressure on living bodies produces an effect which is rarely thought of, namely,—its keeping all the parts about the joints firmly together by an action similar to that on the Magdeburg hemispheres. The broad surfaces of bone forming the knee-joint, for instance, even if not held together by ligaments, could not, while the capsule surrounding the joint remained air-tight, be separated by a force of less than a hundred pounds. . . . . In the loose joint of the shoulder, this support is of greater consequence. . . . In all joints it is the atmospheric pressure which keeps the bones in such steady contact, that they work smoothly and without noise." If there is an unequal pressure on the bones and ligaments, so there must be on the nerves, veins, arteries, and other soft portions of the body, for these are as destitute of internal air, enabling them to resist the outward pressure, as are the bones and joints. It is indeed strange that with bodies so delicately organized, with perceptions so keen that the slightest motion of the air, the mere breath of the breeze fanning the cheek, is distinctly perceived, we should remain unconscious of this thirty thousand pound pressure, or of its continual fluctuations.

We are frequently amused at the effects which are attributed to the variations of atmospheric pressure. The city of Cerro de Pasca is some fourteen thousand feet above the level of the sea. Men, women, and children live in this city, and probably live comfortably there, notwithstanding its elevated site. Cats, however, cannot exist in it; they die in convulsions in a few days after being carried there. And the reason given to account for this fact is, the altitude of the city, and the consequent rarefaction of the atmosphere! Even Humboldt gives this opinion the sanction of his authority, although it is well known that the extent of lungs, and consequent quantity of air received at each inspiration, in any species of animal, is in proportion to the vivacity of its movements, and of course all animals would equally feel the withdrawal of any portion of it. Therefore a diminution of air that would be fatal to cats, would be injurious also to every other animal. Again, it has been stated, that some animals in the mountainous regions of South America, when hunted by dogs, bleed from the mouth and nostrils. This also is imputed to the extreme rarity of the air. Are not fright, fatigue, and accelerated circulation, sufficient to account for the fact, without referring it to the thinness of the air, especially when we remember that over-driven horses at the ordinary level of the earth's surface often exhibit the same symptom?

It is believed by many that the sensations experienced on the summits of very high mountains, are incontrovertible proof of the diminished density of the atmosphere. Unquestionably the *act* of ascending by personal effort to a great altitude, affects the animal organization. It demands an immense amount of muscular exertion, and it is no wonder that the effort is followed by difficulty of respiration, languor, nausea, and other uncomfortable feelings. Instances are given in which blood has gushed from the lips, eyes, and nostrils. Humboldt says, "in the dangerous ascent of the mighty Chimborazo, after *incredible efforts*, at sixteen thousand feet elevation, we began to bleed at the mouth, nose, and eyes." It is not wonderful that, with the full belief that the air was of only half its usual density, this fact should have been imputed to the removal of its accustomed pressure; though it is difficult to say exactly on what principle a *reduced*, yet still equal pressure on the whole body could force the blood from its containing vessels, though it can readily be seen that accelerated circulation from intense exertion might produce the hemorrhage. Many a mouse has been killed by an air-pump without having its blood forced from its mouth; why should the partial vacuum of a great altitude produce this effect? There is no cavity of the human body containing air which has not a communication with the external air, and by the natural channels an equilibrium could be much more easily readjusted, than by the lesion of the blood-vessels.

In looking over Saussure's account of his ascent of Mont Blanc, we came to the conclusion that it was impossible that the painful bodily sensations which he describes should have been produced by the rarity of the air. We observed that twenty men passed the night when near the summit, crowded together in a hole dug in the snow, and covered with a tent-cloth to protect them from the intensity of the cold, thus almost excluding the little air that remained; and that the respiration of the party at the summit, instead of being more full and deep, as if nature would make amends for the thinness of the air by the greater quantity inhaled, was almost suspended. When Saussure's attention was engrossed by his barometrical observations, he almost forgot to breathe; it was an effort, — an act requiring the impulse of the will. He remarks, "all experiments attended with care caused fatigue in this rarefied air, and that *because without thought you hold your breath.*" In the descent, long before they had reached air of effectively increased density, all the painful sensations ceased. It was *the act of descending*, not the level they had reached, that was the restorative. Saussure accounts for the unexpected relief by saying, "As the motion in descending does not press the diaphragm, it does not confine the respiration, and one is not therefore obliged to stop so often to take breath." The party passed the night at a spot only about six hundred feet lower than their resting place of the preceding night, where they had experienced so much suffering. Saussure continues, "We supped with a very good appetite, and I made my observations without any obstruction from indisposition."

It is the circumstance of ascending by one's own effort that creates the bodily distress. We are supported in this opinion by the results of ascents in balloons; for, though a greater altitude has been thus gained than the highest mountain summit ever reached by man, and consequently the air has been according to theory proportionally diminished in quantity, yet the aeronauts suffer no inconvenience on this score. In a former chapter we gave one authority for this fact; we will now quote from Watson's Practice of Medicine. "A diminution of the circumambient weight of the atmosphere is supposed in some cases to bring on hemoptysis. Blood is said to have been forced from sound lungs in persons who have ascended high mountains, where the atmosphere is rare, and where its pressure is sensibly lessened. Perhaps the labor of ascent may have shared in the production of the hemorrhage, for I am not aware that any such effect has ever occurred to persons who have much more *rapidly* reached a very great altitude in balloons."

If the painful sensations complained of were the results of a decreased supply of air for respiration, the discomfort would be experienced by all persons, and under all circumstances, at the same altitude. The strong as well as the weak would feel the loss of air, the bold as well as the timid. The same altitude however differs in its effect on different individuals, and on the same individual at different times. We extract the following passage from Ramond's Ascent of Mont Perdu, the highest summit of the Pyrenees: "I continued two hours on this summit, during which time no being that had life had come within my sight except an eagle, flying with inconceivable rapidity. We breathed without any difficulty in this rarefied air, found by so many so insufficient for respiration. I have been myself more than once or twice witness to persons of hale and vigorous constitutions being obliged to forego proceeding to heights much beneath this. Even Saussure, upon the defile of the

Giant, where the air was by no means so rarefied, experienced an oppression of breathing by somewhat more than common exertion, but here we felt nothing of the kind. . . . So far from occasioning any weakness, it seemed rather to add to my strength and invigorate my spirits." The only effect of the elevation that Ramond specified was a more rapid circulation, indicated by an acceleration of the pulse, and some feverish sensations.

As another confirmation of our views, we will mention that an athletic young friend of ours stated to us, after reading Saussure's account to which we have referred, that he had experienced precisely similar sensations while ascending Baldface Mountain in New Hampshire, which is only about four thousand feet high. He thought several times that it would be impossible for him to reach the summit, so great was his languor and difficulty of respiration. Several times he was obliged to throw himself on the ground in a nearly fainting state. These sensations he thought plainly attributable to the difficulty of ascent, not to the rarity of the air. On Mount Washington, more than two thousand feet higher, the ascent to the summit of which was accomplished chiefly on horseback, the only sensation which he experienced was that of exhilaration.

Another friend has given us an account of two ascents of Mount Washington. He remarked that, the first time, he ascended on foot, and before he had half climbed the mountain, each successive footstep painfully convinced him that the air was becoming more and more rarefied. By the time he reached the summit, respiration was so difficult that he was rather surprised that the change in the quality of the atmosphere should be so sensibly felt. His second ascent was on horseback; this time he experienced none of the former "asthmatic breathing," and at the summit "quite forgot that he was respiring rarefied air."

If the quantity of air is so seriously diminished at great altitudes, how is it that large cities are built at elevations which ought, according to theory, to occasion great suffering to their inhabitants? Among the lofty Andes are many populous cities, enjoying all the luxuries of life, at heights equal to that where Saussure and his companions found it so difficult to breathe on account of the rareness of the air. The silver mines of Potosi are wrought at a height of sixteen thousand and sixty feet above the level of the sea, a height nearly equal to the summit of Mont Blanc; and the city of Potosi stands at an elevation of thirteen thousand three hundred and fifty feet on the declivity of the mountain. The city of Quito, with a population of seventy thousand, is nine thousand feet above the sea. Among the Highlands of Thibet also are cities standing at great altitudes, and surrounded with such luxurious vegetation, that Chinese writers, it is said, describe them as regions of pleasure. In this connection we will mention that a series of meteorological observations, continued for twenty-four hours in succession, has recently been made on one of the summits of this range, at the height of eighteen thousand four hundred feet, where, according to the barometer, the air was of less than half its usual density, and "the chief result was, that the curves followed very nearly the same changes that they were observed to do in the lower regions."

Vegetation does not appear to be affected by altitude in any other manner than as the height diminishes the mean temperature. Different species of plants become dwarfed, and then disappear as the elevation increases, just as they become dwarfed and disappear as we proceed from the equator towards either pole. In central Europe the decrease of heat is 1° 8′ for every five hundred and thirty-four feet of vertical elevation; an ascent, therefore, of two hundred and fifty feet is equivalent to a progress of about one degree of latitude. Alpine plants are commonly covered with a thick and close down, which, as has been remarked, is analogous to the soft fur of northern animals. This Alpine vegetation occurs at regularly decreasing elevations as we advance northward, and in the Arctic regions it is the vegetation of the plains, where the temperature is the same as at the lofty summits of the equatorial zone. The plants of Alpine heights have been successfully cultivated on the plains below, when due care has been taken to give them the peculiar treatment required.

But it is supposed that the great difference of atmospheric pressure *ought* to affect vegetation, and consequently a difference between plants indigenous at different levels has been traced out. "The influence of atmospheric pressure," says a distinguished naturalist, "seems to me particularly evidenced in the great, I may say the prevailing number of Alpine species endowed with a volatile fragrance, which adds so much to the sweet and soothing influence of mountain rambles; whilst the northern species, however similar to those of the Alps, partake more or less of the dulness of the heavy sky under which they flourish." He adds, however, "It would be a mistake to ascribe to reduced atmospheric pressure the peculiar aspect of most plants in the higher Alps." May we not consider this fanciful proof of the pressure of the atmosphere, rather as an expression of the poetical idea, that science often values the bouquet in proportion to the labor of gathering it, and that, as men ascend above the defilements of earth, the sweeter is the aroma which nature sheds in their path?

As we have before said, Nature, so far from hiding from view, throws herself open to the sight. By change in the mode of her action she appears to invite inspection, — by presenting herself under many aspects she seems to court our examination. Were the earth but one unvarying plain, many of her processes which are now fully exposed, would remain hidden. The mountain transports the northern vegetation to the tropics, causing the polar plant to send forth its fragrance under the rays of the vertical sun; it throws up the strata of the earth, as it were for examination; it collects together the extremes of temperature; it places side by side the burning lava and the never-thawing snow, — thus condensing world-wide phenomena, that their relations one to another may be perceived at once. It displays not only physical results, but the forces which control them, showing us in miniature the results of the variations of rotary force, the successions of climate, and the changes of electric intensity. Every fact therefore connected with the "cloud-capped" mountain, becomes full of interest to all who seek to understand the organization of the world, and to trace out the laws which govern its elements, — none more so than the enveloping wreath of vapor which has furnished its poetic epithet.

"In the tropical region of Paramos," says Humboldt, "at an elevation of between twelve and fourteen thousand feet, some species of large-flowering, myrtle-leaved Alpine shrubs are almost constantly bathed in moisture, affording evidence of the frequency of aqueous precipitation, as do the frequent mists with which the lovely plateau of Bogota is covered. Mists arise and disappear several times in the course of an hour in such elevations as these, and with a calm state of the atmosphere. These rapid alternations characterize the elevated plains of the chains of the Andes." By this condensation is furnished from the humid atmosphere the supply of force required by the high level of the mountain, the moisture being brought to it from the beds of the ocean, which we have described as the storehouse of force. Every consideration is interesting which brings to our thoughts that mighty Power, which, while it supplies the wants of the inorganic world, provides also its banquet for the wayfaring sparrow.

To return from what may appear a digression, — we omit, for want of space, a physiological view of our subject which had been prepared; but we will observe that it is at once apparent, that the act of ascending exhausts the strength, the animal force, much more than the same quantity of muscular movement at one level. The ascent of a single flight of stairs is felt by the weak, and the strongest experience some difficulty of respiration in rapidly mounting several flights in succession. If we call to mind the fact, that the ascent of a mountain is the act of lifting the weight of the body to the whole extent of the elevation, — that it requires the rotary force for the new level to be supplied, — we shall not be at a loss for the reason of accelerated respiration, exhaustion, and of many other disturbances of the animal functions. Besides these, however, there may be a class of sensations experienced at a very great altitude, the cause of which is independent of the muscular effort of the ascent.

We know that altitude causes a difference in the electric condition of the atmosphere; positive electricity normally increases in exact proportion to height. To the usual degree of electricity the body becomes habituated, and an increase or diminution of the accustomed quantity always affects the feelings. If it is lessened, there is a depression of the animal spirits, while a slight increase brings with it a feeling of new vigor accompanied by exhilaration. Contrast the sensations experienced in a warm, damp summer-day, with those of a clear, cold day in mid-winter. On the other hand, a greater amount enervates the body by a feverish increase of the circulation, and invariably diminishes the activity of the respiration. We know from experience that an unusual electrical excitement of the body has the effect to enervate the body, flush the cheeks, and almost to suspend the breath.

The electricity of the atmosphere, by its fluctuations at the surface of the earth, by its changes according to elevation, and its differences according to the seasons, in its silent unobserved transfers from one element to another, or in the sudden loud explosions of the thunder-storm, exercises a powerful influence on the vegetable and animal organization. By tracing its effects, not only on that which has life, but also on unorganized matter, we shall be able to solve many of the mysteries of nature. How much positive enjoyment might be received, how much uneasiness and illness might be avoided, by a recognition of the power over the body, exercised by the electrical states of the elements about us! But this will never be, so long as gravitation is regarded as the controlling cause of even bodily sensations. In a damp day, when moisture mingled with the air strips the body of its accustomed degree of electric force, and depression of spirits ensues, we look at the sluggish vapor resting on the surface of the ground, and consider it a proof that the weight of the air, its pressure upon the body, is diminished. So, when the cheek is flushed, and the blood leaps through the veins with exhilarating power on the mountain height, we again trace this opposite effect to the lessened pressure of the air. And again, when by the toil of ascent the respiration is hurried, and the body sinks with fatigue, still no cause is sought for all these diverse effects, but the variations in the space which the atmospheric particles occupy!

We would mention another instance of the tendency to impute otherwise unexplained phenomena to alterations in the pressure of the air. It is well known that the diurnal variations of the barometer are in some way connected with the diurnal variations in the intensity of the magnetic force as indicated by the needle. There are every day two maxima and two minima of each, — the maxima of one corresponding to the minima of the other. There is also a diurnal variation in the tension of vapor in the atmosphere; and, as vapor is believed to change the weight of the air, this is supposed to account for the oscillations of the barometer. Of course then the variation in the magnetic power is also traced to the state of the atmosphere. Thus all investigation seems to stay itself upon the condition of the air, when further research most unquestionably will show that the variations of the barometer, and of the magnetic force, and the different degrees of the tension of vapor, are traceable to one cause, — to the changing position of the surface of the earth with reference to the centre of her movement, — to the transfers of force which preserve the equilibrium of her motion. These changes, as extensive as the surface of the earth, corresponding to her daily movements, varying too with the seasons as she traverses the different parts of her orbit, — must they not all depend upon the same cause, — the force which controls her daily rotation and her annual revolution? The same force which circulates through the earth permeates the air, and in its changes and transfers this force must move the barometer, touch the needle, and determine the rise and fall of vapor.

We asked, in a former chapter, why continue to trace all the phenomena of this beautiful world to one cause only, and that a cause, allowing to it all the virtue and efficacy which it can reasonably claim, so inadequate for their explanation, its only legitimate operation being to determine the position of matter, — the point of space which any atom should occupy? Does it not repress inquiry, and arrest the mind in its search after truth? The belief in the attractive force of matter is not only the foundation of the structure of our present philosophy, but it places every stone of the edifice, and superintends even its finish and its ornament. The mind has no occasion to exert itself further. By gravitation is explained the act of drawing the first nourishment, and the last expiring breath; the child totters under it in its first attempts to walk, and old age resists it with its staff. We see only this force of senseless matter when the pendulum swings, when the stone falls, when the vessel floats, when the vapor rises, when the clouds hang heavy over our heads. We trace its effect in the descending rain, in the rushing of the mountain brook, in the flow of rivers, in the swell of the tides, in the outline of continents, in the shape of the earth, in the form of her orbit, in the rush of the solar system through space. We imagine even that by the whirling of "star-dust" new worlds are gradually added to the realm under the sway of the attractive power of inert matter!

We pause for one moment to glance back at the extent of ground over which we have so rapidly passed. The facts that we have endeavored to examine are numerous and varied in their character. Had we the ability, we have not the space to present them in their fulness of scientific detail, nor is it necessary in a treatise of this nature. But we have attempted in simple language, with the least possible use of technical phraseology, to group together many phenomena under a new theory for their explanation. The result, in spite of errors and mistakes, we hope may tend in some degree to strengthen interest in the works of nature; for in whatever light the world is viewed, in whatever aspect it is considered, its beauty is apparent; and though the offered theory be rejected, it may benefit some minds to forget for a moment the established explanations of things, that they may see creation with a light from a new direction touching some of its prominent points. If we have substituted for logical deduction an allusion to the "laughter" of the exulting wave, — if in place of the formula of the mathematician we have introduced a text of Scripture, — if we have questioned the attractive power of matter, and doubted whether it be indeed the cause of all the varying phenomena about us, — if we refuse to admit the idea that the universe exists in its present form from the necessity of the case, because it could not be otherwise "considering the known properties of matter," — it is because when we look at nature, we wish to see more than the mere external lineaments; — we would trace out the "Indwelling Spirit" which pervades every part; which is the cause of all change, of all movement; which is to the universe of matter what the life of man is to the body with which he is clothed. We would *feel*

"A presence that disturbs us with the joy
Of elevated thoughts; a sense sublime
Of something far more deeply interfused,
Whose dwelling is the light of setting suns,
And the round ocean, and the living air,
And the blue sky."

CHAPTER IX.

"THOUGH IT BE ONE OF THE MAXIMS OF THE TRUE PHILOSOPHY, NEVER TO SHRINK FROM A DOCTRINE WHICH HAS EVIDENCE ON ITS SIDE, IT IS ANOTHER MAXIM, EQUALLY ESSENTIAL TO IT, NEVER TO HARBOR ANY DOCTRINE WHEN THIS EVIDENCE IS WANTING."

Dr. Chalmers.

THE tides are often referred to as a proof that the law of gravitation extends to the spheres. They are looked upon as the visible and tangible results of the attractive power of the sun and moon over the waters of the ocean, — as a fact which appeals to the senses, — as an evidence of the doctrine adapted to the popular mind. Without the tides, it is said, "the principle of the universal attraction of matter would rest solely on abstruse reasoning, on abstract mathematical deduction." The tides indeed are presented by many, as settling the fact, that the law of falling bodies extends upward to the spheres, giving them their motions and their mutual influences.

But the tides fail altogether to answer this purpose among the unscientific. It is popularly believed that the sun and moon are, in some way or other, connected with the phenomena; how, and why, is not understood. There are questions asked which the books do not answer; doubts entertained, which the theory of gravitation does not dissipate. There is rather an acquiescence in the statements on this subject, than a full and hearty assent. There is, too, a wide-spread rumor that philosophy itself, with all its strength, and with all its help from abstruse mathematical reasoning, is baffled in the attempt to explain the tides as a result of the law of gravitation.

We believe that, if the sun and moon do attract the waters of the earth, this attraction cannot produce the phenomena of the tides. To strengthen us in this belief we have high authority.

Professor Brande expresses the opinion that the collected facts on this subject are in advance of the theory; and that these facts wait for a more accurate determination of the principles of hydrodynamics, or the laws of the motion of water. La Place expressed the belief that the collection of more facts, at the observatory at Brest, a place well adapted for observation, would account for the discrepancy of the theory with the then established facts. Whewell, in the Philosophy of the Inductive Sciences, says, "The tides are a portion of astronomy, for the Newtonian theory asserts these curious phenomena to be the result of the attraction of the sun and moon. Nor can there be any doubt but this is true as a general statement, yet the subject to the present time is a blot upon the perfection of the theory of universal gravitation, for we are very far from being able in this, as in other parts of astronomy, to show that theory will account for the time, magnitude, and all the other circumstances of the phenomena at every place on the earth's surface. . . . . And what is the portion of our mathematics which is connected with this solitary, signal defeat of astronomy? The attempts of the greatest mathematicians, Newton, Maclaurin, Bernouilli, and La Place, to master such questions, all involve some gratuitous assumption, which is introduced because the problem cannot be otherwise mathematically dealt with; these assumptions confessedly render the result defective, and how defective it is hard to say."

Yet notwithstanding that the theory is thus questioned by truly great men, it is still presented to the popular mind as a proof of the law of universal gravitation. The assumption that this "blot" arises from defects in mathematical reasoning, or from ignorance of the laws of the motion of fluids, is entirely gratuitous; for, without recourse to mathematics, without the assumption of any laws of hydrodynamics except those universally acknowledged, the discrepancy between the facts and the theory is fully apparent.

We would call attention in the first instance to this incontrovertible position. The power of attraction, according to the law of gravitation, and consequently the attraction of any heavenly body, is inversely as the distance, and directly as the quantity of matter; or, popularly, attraction decreases with the distance, and increases with the quantity of matter. If this law is abandoned, the theory of gravitation is annihilated at once; if it be retained, the theory of the tides must be abandoned. For the waters at the surface of the earth are nearest to the earth, and the quantity of matter of the earth is as forty-eight to one, compared with the quantity of matter composing the moon. Now with this vastly inferior attractive power, how can the moon's attraction overcome that of the earth? How can the moon raise the water in the least degree? A magnet of inferior power never detaches a piece of iron from a magnet of greater attractive force.

Again, if the moon can raise the waters of the ocean from the earth, the attraction of the earth decreases as the waters go up from her centre, while the attraction of the moon increases as they come nearer to her centre. The idea, then, that the moon can elevate the tidal wave one inch, includes the idea that the moon can draw the waters of the earth to her very surface, — the grasp of the earth being gradually relaxed, and the hold of the moon gradually strengthened as the waters ascend.

Or else there is a focus or common centre somewhere between the moon and the earth, where the gravitating forces of the two spheres produce an equilibrium, and where the elevated waters would remain, and be at rest, — the property neither of the moon nor of the earth, — belonging exactly to neither sphere.

If the moon attract the waters of this globe, then she attracts equally all the water of one hemisphere, the hemisphere nearest to her, and depresses equally all the waters of the other hemisphere. We say equally, for it is so practically; theoretically the waters at the point nearest the moon would be attracted more than the waters at the horizon in the proportion 244.000^2 to 240.000^2. The result of this would be the equal elevation of all the waters of one hemisphere, and the equal depression of the waters of the opposite hemisphere; and this inequality of the height of the waters would be propagated round the earth by her rotation. The result of this attraction by which the earth is forced to relax her hold on a portion of her waters, and increase the hold on other portions, would be a current of water rushing round the globe at the rate of one thousand miles an hour.

The comparative power of the sun and moon as mathematically deduced from their different quantities of matter and distances, to serve for the theory of the tides, is roughly stated as twenty to forty-nine. The sun, then, has less than half the attractive power of the moon at the surface of this globe. If the moon has this superior attraction for one purpose, she has it invariably for all purposes, and should manifest this superiority not only in the phenomena of the tides, but in every phenomenon connected with the gravitating or attractive power. The glorious luminary, the centre of attraction of all the planets, so large that if they were united in one mass, their aggregate would be to the sun as a pebble to a mill-stone, stands aside shorn of his power, degraded to the position of a satellite to the moon,—his humble office, in the words of the theory, being "to prime and lag" the tidal wave which follows the sweep of the majestic moon across the horizon. The sun, when drawing apart with a force of 355,000 spheres of the size of the earth, manifests not even the power to uplift a little tidal wave, to follow in the wake of the enormous wave of the queen of the skies, the solar tidal wave which, if he have any power, should at times follow the lunar wave easterly, and then again westerly being never seen; for even this little display of force the theory cannot allow him because of stubborn facts.

But let those who uphold the theory proclaim its defects for themselves. It needs not our aid to show that it is utterly insufficient, and that it is an induction from the theory of gravitation, not a deduction from the phenomena which the tides present.

We make the following quotations from the article on tides in the London Encyclopedia: "The reader will probably be making some comparison in his own mind of the deductions of this theory with the actual state of things; he will find some considerable resemblances, *but he will also find such great differences as will make him doubtful of its justice.* In very few places does the high water happen within three fourths of an hour of the moon's southing as the theory leads him to expect, and in no place whatever does the spring tide fall on the day of full and new moon, nor the neap tide on the day of her quadrature. These always happen two or three days later." After referring to some other deviations of fact from theory, which deviations the writer accounts for by the operation of local causes, he continues, — "There is also a general deviation of the theory from the real series of the tides. When the moon enters her second quarter at noon, it is high water at Brest at 8.40 instead of 9.48, which theory assigns. Something similar and within a few minutes is observed at every place on the sea-coast. This is therefore general, and indicates a real defect in the theory. Indeed, we have no rules but what are purely empyrical, or which suppose a uniform progression of the tides."

The theory of the tides could not stand for a moment were it not held up by the doctrine of *vis inertiæ*. The indistinctness of this *vis inertiæ* is a very convenient mist to throw over the defects of the theory. If the flow of water does not show itself at the right time, — at the time fixed by theory, *vis inertiæ* of rest holds the waters back; on the other hand, if the flow arrives uninvited by the theory, *vis inertiæ* of motion brings the unwelcome presence. This *vis inertiæ* has been called "a force of inactivity, — a forceless force, analogous to a black white, a cold heat, and a tempestuous calm. For these reasons we are inclined to think," continues our author, "that bodies are wholly passive, that they endeavor nothing, that they continue in motion or in rest by whatever begins the motion, and continue to move so long as the cause continues." But the property of *vis inertiæ*, as we have said, is convenient when stubborn facts will not bow to the asserted theory. Thus the happening of spring tides two or three days after the full or new moon can be accounted for, — the tidal wave continuing to rise after the cause of its rise is weakened, from the habit which the water has acquired of increasing its flow.

We copy the following explanation of one of the doubtful points in the present theory of the tides, from Sir David Brewster's Life of Newton. "But the most perplexing phenomenon of the tides, and one which is yet a stumbling-block to persons slightly acquainted with the theory of attraction, is the existence of high water on the side furthest from the moon at the same time as on the side next to the moon. To maintain that the attraction of the moon draws towards herself and at the same time from herself, seems at first sight paradoxical. But the difficulty vanishes when we consider the earth, and the waters on each side of it as three distinct bodies placed at different distances from the moon, and of course attracted with forces inversely proportioned to the squares of their distances; the waters nearest the moon will be much more powerfully attracted than the earth, and the earth more than the waters farthest from the moon. The consequence must be that the waters nearest the moon will rise from their level, while the earth will be drawn away from the waters behind the earth, which waters as it were will be left behind, and be in the same situation as if raised from the earth in a direction opposite to that in which they were attracted by the moon." Sir John Herschell explains the difficulty in a similar manner. Indeed, it is probably the best solution that can be given under the theory; and it passes current without examination, notwithstanding its obvious defects. One would suppose that the earth's attraction might keep the waters at the side opposite to the moon close to her bosom, especially when assisted by the attraction of the moon drawing in the same direction.

Gravitation, if we admit all that is claimed for it, will not cover the phenomena of the tides. Imagine a miniature world, with its dry land, its lofty mountains, its wide-spread oceans, and its enveloping atmosphere, placed within the sphere of the earth's attraction. How would the earth act upon this little world? She would draw it towards herself, with a force operating equally on its every atom. If it moved toward the earth through a resisting medium, the more solid parts might advance more rapidly than the rarer, — the granite rocks than the thin air; if it moved through void space, every part would advance with equal rapidity, all retaining the same relative position.

So if the earth is attracted to the moon, every atom is attracted; and, if there is any motion induced, every part moves equally. Indeed, astronomers assert that "the earth *is* constantly falling to the moon, being continually drawn by it out of its path, the nearer parts more, and the remoter less so than the central." But the attraction of all the particles of the earth for each other, which gives it its form, — which moulds it into "this goodly ball," is too great for the moon to separate its elements, to tear it apart as it were, — to elongate its mass so as to cause the tidal flow. If the moon attracts the earth at all, it must be as a whole, — as one mass. She cannot have greater power over one portion than over another of the earth's material. The theory of gravitation does not give the moon elective attraction, by which she is enabled to move the waters of the ocean, yet leave untouched the waters of the great lakes, or to lift the ocean and leave unmoved the bed on which it rests. And if she can lift water away from its own centre of attraction, why does she not raise the sands of the desert, the atmosphere, vapor, and the descending rain? So evident is this idea, that a scientific man in France, after writing a treatise on the tides in accordance with the present theory, remarked, how strange it is that the moon should possess this power over the waters of the ocean, when we know she cannot draw towards herself a feather floating above them!

The extension of the law which is supposed to determine the fall of bodies, into a universal principle pertaining to all matter, is considered as one of the most important achievements of philosophy. The great glory of the theory is its universality, — the extent of ground which it covers. The tendency of philosophy therefore is, never to narrow its range, but by all possible means to extend its application. A fact that has once passed under its sway is held with a tenacious grasp. It is withdrawn from farther examination.

The theory of the tides is an instance of this. The tidal wave rises up against the very law of falling bodies; it escapes from the attraction of the earth. Looking upward, astronomy sees the moon, a smaller body than the earth, at a great distance from it, and endows her with power to overcome the attraction of the greater earth, as regards the waters lying upon her very bosom. The mathematician cannot reconcile the theory with facts, the scientific man looks incredulous when he teaches the doctrine, and plain common sense absolutely refuses to believe it. Thus it remains a doctrine assented to, but not believed, — asserted without being fully comprehended, — declared by men of the strongest minds to be a "*blot*" upon the theory of universal gravitation. Why not boldly abandon the theory, and seek for a better? Facts will not support it, and facts never change their argument. They will not do in the future what they have refused to do in the past.

Our knowledge of the great tidal waves of the ocean is gained only through the facts presented by the tidal ebb and flow. The swelling and shrinking of the ocean is too small compared with the extent of its surface to be perceptible; for above its vast expanse there is no fixed point by which to measure its altitude. But on the coasts the consequent currents are noticed, and the rise and fall of the waters can be measured. Only by these changes of level on the shores, are the alternations of the great deep made known. It is the tidal ebb and flow which indicate the character of the tidal wave producing them.

The ebb and flow both continue for six hours, that is, from elevation to elevation, and from depression to depression, there is an interval of twelve hours. This, however, would bring the time of high water to the same hour every day; but in addition to the regular interval, a half rotation of the globe, there is a lapse of time between the tides, occasioning what is called their progression, which continually retards the time of high water to that degree that the progression completes its period, and the phases of the tides complete their revolution, returning to the same hour, in one lunation, or in the time of one revolution of the moon. The tide commences its flow sluggishly, gradually increasing its velocity until towards high water; it then slackens its speed, and when the full height is attained, it pauses awhile until the depression of the great tidal wave recalls the waters, which in their ebb manifest the same increase, diminution, and cessation of current.

These are the daily phenomena of the tides, the extent of their flow and the direction of the currents being somewhat modified by local and temporary causes, such as the configuration of the coast, and the force and direction of the winds. Besides these irregular and occasional fluctuations, there are also periodical changes which are important, as they point to the nature and character of the producing cause of the tidal wave. These periodical variations are governed by the seasons of the year, by the position of the moon in relation to the earth, and by the hour of the day at which high water occurs. The greatest or spring tides take place two or three days after the new and full moon, and the smallest or neap tides at about the same lapse of time from the quadratures. There is also a difference between the day tides and the night tides, and between those of summer and of winter.

To obtain a distinct view of this subject we must separate the effect from the cause, — the current running to and from the coast, increasing and diminishing the waters in bays and harbors, from the elevation and depression of the ocean; we must distinguish the tidal flow from the tidal wave; for they are separate and distinct in their character. The great tidal wave is the vertical elevation of the waters of an ocean, — the tidal flow is merely a surface current, running on an inclined plane, at the rate of a few miles an hour; and as the tidal wave by its rise or fall changes the inclination of the plane, the consequent current changes the direction of its flow. The tidal flow therefore is the result, not the cause, of the alternations in the level of the ocean. Compared with the great oceanic currents, the immense westerly current of the tropics and its branches, and the polar currents, the tidal flow is utterly insignificant. But all these mighty ocean currents are lifted up with the great tidal waves.

To form these waves the ocean waters of the two opposite hemispheres are elevated, and there is an accompanying depression at the equidistant points; and by the rotation of the world each part of its surface shares in the alternations. The question is, what is the impulse of this vast oscillation,—what the dynamic power for this alternate rise and fall of the waters of the globe? The even-balanced sea, though so free to move, does not of itself lift and depress its volume. There is a moving power,—an impulse which gives the mighty motion. As winds in passing over the surface of the sea form minor waves, so is there some force which penetrates the waters, stirring them to this vast tidal pulsation, swelling and shrinking them on every shore, in the bays and harbors of every coast. So that instead of a dead level, or a level storm-disturbed only, there are ever varying and periodic alternations of surface, ministering to the wants of man, and presenting to him one of the most interesting phases of the "life of the world."

In all oscillatory movement, as in the swing of a pendulum, there is a transfer of force according as the level changes. The descending water gives out the no longer needed force of rotation to the water rising to the higher level, the mean altitude and the mean degree of force remaining ever the same. But oscillation does not furnish its own impulse. To account for the tidal wave, we must recur to a regular and periodic passage of force which touches the waters, imparting to them their motion.

The force which causes the tidal swell must be of varying intensity. We refer not to local and occasional changes, but to periodic alternations in its strength. The motive power of the tides varies in degree according to the position of the part of the earth at which it is high water in relation to the centre of her orbit; it varies according to the position which the earth occupies in her orbit, and her position in relation to the moon. In other words, the hour of the day, the season of the year, and the phases of the moon point to the changes in the intensity of the impelling force of the tidal oscillation. It is then connected with the earth's motion; it is a branch of the same force. It comes therefore from that which is world-wide in its action, from that which increases and diminishes the great movement of the earth; for we trace all its changes to the perturbations and vicissitudes of her rotation and revolution.

The force of terrestrial magnetism was formerly supposed to be of limited and partial action,—to pertain to a class of metals only; but recent discoveries prove the universality of its power over every kind of matter of which the world is composed. The oxygen of the atmosphere has been recently proved to be the subject of its power; and certain crystals suspended by a thread of silk are found "to take under the direction of the earth's magnetism a determinate and fixed direction," acting as magnetic needles, "pointing constantly towards the poles of the earth, towards the magnetic poles, or toward some azimuthal point." The transfers of electricity, once considered as the passing of matter from the positive to the negative electric state, are now regarded as the plus and minus in the present quantity of a pervading power, and electrical phenomena are deemed the evidence of a restoration of the equilibrium of force. Philosophy recognizes a force flashing through the sea, permeating the atmosphere, and penetrating into the solid earth.

Nor is this force spasmodic and irregular in its transits,

There are permanent currents indicated by the magnetic needle; for the constant action of the needle can be caused only by a continued flow of the galvanic stream; nor is the flow of uniform intensity, but it is subject to regular and periodic alternations of strength. Between the tropics, where by the greater diameter, and the consequently greater degree of rotary force, the electrical and magnetic phenomena are more fully developed, the hour of the day can be determined "by the position of the needle, as well as by the changes of the barometer." Humboldt believes in the existence of many systems of magnetic currents; he describes the incessant oscillation of all magnetic phenomena "according to the hours of day and night, and according to the seasons of the year, and according to the whole course of the year," so that we are compelled to believe, that there are regular and constant transmissions of force around the world, to preserve the equilibrium of motion between the different parts, as by their changing position there is a necessary acceleration or retardation of their velocity.

In our remarks on the barometer we referred to the oscillation of the earth's circumference, by which there was at the opposite points an enlargement of the diameter, and a contraction at the intermediate, equidistant points. It is not the barometer only which indicates this fact. The action of the needle strengthens us in this opinion. If there is this swelling and shrinking, there are parts of the globe which, by their wider orbit of rotation, would require the greater force of propulsion, and parts which by their subsidence would require the less. There must needs be a transfer or passage of force, and, the action of the needle being determined by the current of force, this transfer would be indicated by its movements. Accordingly we find, that the needle has its two maxima and minima of horizontal intensity, corresponding to the maxima and minima of the barometer, coinciding with noon and midnight, morning and evening. In the transfers of force to preserve the equilibrium of motion, the current would be the strongest and the weakest at the points indicated by the corresponding degrees of magnetic intensity; at those parts the barometer indicates the greatest change of level, which causes this transfer of the force of rotation.

Again; there is another manifestation of the need of more and of less force at these opposite points of the earth, from the differing degrees of the intensity of the tension of vapor. Without any known cause, water at certain times passes into vapor, and at others the vapor condenses under the same degree of temperature. Besides the occasional and irregular exhibitions of this phenomenon, there are also periodic alternations, corresponding with the hour of the day or night. The tension of vapor has also its two maxima and minima, corresponding to those of the barometer and the compass. We believe that this alternation is governed by the need of more or less force of rotation of the several parts of the earth. When force is required the vapor condenses, — when it is super-supplied, the water passes into vapor. We have repeatedly referred to this as one of the means by which the equilibrium of motion is maintained. At first sight, the influence of this alternate vaporization and condensation may appear of little value; but on the great scale of nature's works, with the wide expanse of the surface waters of our globe, it is indeed a vast agency. It has been supposed that vaporization and condensation were determined directly or indirectly by temperature; but how can this be when vapor is formed alike under the vertical sun of the tropics and from the frozen oceans of the poles, snow and ice, as well as the warmest surface water, throwing off vapor which condenses at times in all altitudes, and under every degree of sensible heat.

The temperature, positive or relative, produced by the sun's rays, appears altogether insufficient as the direct or indirect cause of the passage of the currents of force. It cannot explain the maxima and minima of the barometer, of the needle and of the tension of vapor, any more satisfactorily than it could explain the two daily maxima and minima of the tides. The sun's influence changes the temperature of the earth only a few feet beneath its surface, and the temperature varies according to elevation; yet the barometer, whether suspended over sunny plains or carried to the summits of snow-clad mountains, notes equally the diurnal changes; and the needle in caves or mines, where the thermometer shows that the solar influence never penetrates, records the hourly increase or diminution of the magnetic stream by which its motions are governed. Magnetic storms also pass over vast portions of the earth, disturbing the regular action of the magnetic needle, whatever may be its altitude of position or the surrounding temperature; and if the needle is affected all over the world thus simultaneously by magnetic storms, surely that which causes its regular action has an equal range, — a range far too vast to owe its power to changes of temperature, which from their nature are limited and local. If the perturbations of the needle are world-wide, so is the cause of its normal action.

Unquestionably all these phenomena, constituting the threefold manifestation of the transfer of force, are connected with heat, — not with heat from solar influence, but with the sensible heat evolved by the passage or readjustment of the magnetic force. The temperature of soft iron is raised by the process of infusing into it the magnetic power; the heat of the earth, below the surface at least, we believe is determined by its degree of magnetic force; the magnetic poles are the poles of extreme cold. Transfers cannot take place without indicating that heat is another term for the force of motion. "Terrestrial magnetism," says Humboldt, "stands in most intimate relation with the internal as well as the external heat of our planet. . . . But the old explanation of the horary variations by the progressive warming of the earth by the rotation of the world, must be limited to its upper surface." For the solar heat and the consequent changes of temperature do not always reach the instruments which alternate in their action. They are touched and moved by the force of terrestrial magnetism, and their variations record the changes in the intensity of its passing currents.

But, besides the transfer of force for the adjustment of the rotary motion of the parts of the earth, which by the oscillation of surface need the greater or the less degree, there is another flow of the magnetic power to restore the equilibrium of the earth's motion in her orbit of revolution. They both affect the needle, combining to produce its varied phenomena as well as the many meteorological changes and vicissitudes, which present the action of the two causes in various degrees of relative intensity. This globe requires less of the force which impels her in her orbit at some parts than at others. The point nearest the centre of revolution describes an orbit eight thousand miles less in diameter than the point which is the most remote, and generally one hemisphere needs less of the force of motion than the other. The rotation of the earth continually changes the position of every part in relation to the centre of the orbit. There is, therefore, a constant, unremitting passage of force from one part to the other, from one hemisphere to the opposite. This which for the want of a better name we call the *spare force of revolution*, we consider as the motive power of the tides, — as the impulse which causes the oscillation of the great deep.

That there is this passage of electrical or magnetic force around the globe is acknowledged; and it seems evident that it cannot be caused by the progressive warming of the surface of the earth. If we are right in our conceptions of the nature of force, a flow must be produced by the unequal orbits described by the different parts of the earth, as they are more or less distant from the centre of revolution. This inequality appears to us sufficient to produce a transfer that would account for the results which we impute to it; for there are instruments of such delicacy, as to detect a variation of one forty-thousandth part of the magnetic intensity, and surely nature would register in her phenomena the differences in the degree of required force in orbits which vary eight thousand miles in diameter.

There is another consideration which supports our belief in the reality of this transfer of force. It is, if we may be allowed the expression, the necessity of a life-giving circulation, to impart to the face of nature its ever-varying beauty, to produce those changes and vicissitudes which minister to the enjoyment of life, and to create that ceaseless activity of the elements which brings the world into sympathy with the activity of living beings. Without this it would seem as if all things would harden into monotony, and the power now so fully exhibited in exciting and controlling the elements, would be hidden by the dull, unvarying uniformity of its action. We live in a world of change, but of change so limited that the storm only heightens the enjoyment of the calm, while it has power and beauty of its own, — of change so limited, that instead of manifesting chance, it unfolds design, the cycle of vicissitudes showing the steadfastness of the Hand which governs them.

Our belief in the necessity of this constant transfer of force is further supported by considerations drawn from the arrangements, by which activity of motion is produced in the other planets belonging to the solar system. The strength of this flow of force is governed, not only by the diameter of the sphere, but also by the proportion which this diameter bears to the diameter of the orbit. That each world may have a sufficient current of force circulating through its several parts to produce the necessary alternations, and yet these alternations be kept within such bounds, that life and the enjoyment of life may be secure, the diameters of the planets increase as their orbits enlarge, and by the addition of satellites, and by the increased number of these attendant spheres, as well as by the wider sweep of their orbits, the extreme points of each local system are more widely separated as their distance from the centre increases, and an equality in the transfers of force is maintained among them all. This is the case with all the planets, except the system of smaller spheres between the Earth and Jupiter. The Asteroids have a greater eccentricity than the other planets, and by the greater inclination of their orbits revolve in a different plane, which may perhaps stand in the place of the greater diameter. Were Jupiter placed next the Sun, the wide distance between its extreme parts would bear too great a proportion to its distance from the centre of its orbit. The different degrees of force needed by the parts at the outer and inner circumference of the orbit, would cause too rapid transfers; while Saturn, at his distance, can bear the enlargement of diameter caused by his rings; and perhaps in far-off Uranus the retrograde movements of his satellites gives the required inequality. Nor can we fail to remember, that the outermost planets have the greatest velocity of rotation.

Returning to the consideration of our own planet, we find not only the transfer of force which is due to her diameter, but also those variations of its intensity which proceed from eccentricity of orbit, and from the action of her satellite. The motion of the earth in her orbit is not of uniform velocity, nor does her distance from the sun always remain the same. Hence the increase and diminution of the transfer of force is dependent on the season of the year; hence has the needle its two annual maxima and minima; hence the yearly barometrical changes, and the different degrees of the tension of vapor. The maxima and minima of the year correspond to summer and winter, spring and autumn, as those of the day to noon and midnight, and morning and evening. Thus "all things are double one against another, and He hath made nothing imperfect. One thing establisheth the good of another."

There can be no independent change of motion in any mass or any sphere. The force which one receives another surrenders. All things therefore move sympathetically,— there is no isolation in any part of the universe. The spheres move in unison, as if connected, (as the idea has been expressed by another,) by elastic material ties. Disbelieving altogether in the attractive and repulsive power of matter, we regard the universe as one great whole, its separation into parts in man's mind being only because he must divide in order that he may comprehend.

If the heavenly bodies generally move thus in sympathy, if the vibration of the more distant stirs the force that gives the complementary oscillation to the others, — how much greater the mutual influence between planets and their satellites, whose relations seem designed to increase and to vary the strength of the life-giving flow of force!

The relation of the moon to the earth is not thus intimate because of her nearness of position only. We regard the moon as a part of, as pertaining to, this globe; we consider them as together constituting one member of the solar system, revolving together around their common centre. The moon is not an independent mass wrapped up in herself by an individual rotation, as if a separate value were included within her own diameter. Her secondary movement is round the earth as her centre. It is true, as astronomy asserts, that, in passing round the earth, she turns in relation to a point in space. So does every mountain on the face of the earth. Mont Blanc turns in relation to space once for every diurnal revolution of the earth, otherwise its base would not rest on the earth, and its summit point continually from it. So the moon, because she turns in relation to space, keeps, — her libration excepted, — the same hemisphere always directed towards this globe. We repeat the idea; she does not by an independent rotation wrap up an independent value in her diameter.

Besides, in the associated motion of the earth and moon around their common centre, there is between them a point of equilibrium which describes the true ellipse of their common orbit. This point of space, which would be the true position were the masses united, is near the actual line of the motion of both at the moon's quadratures; at the new and full moon they are both at the greatest distance from the mean orbit. Hence come the monthly periodical changes of the earth's motion, and with these changes the periodical increase and diminution of the flow of force, — the increased or diminished spare force of revolution, which records itself in the greater or less flow of the tides, and in the host of connected phenomena.

The maxima and minima of the tides take place after the new and full moon, and after the quadratures. If the cause of the increase and diminution of the current of force is the oscillatory motion of the earth in her orbit, the onward movement of the vibration would continue after the cause had passed its maximum; the motion of the earth would go on, and its greatest aberration would be some two or three days after the culminating point of the satellite which determined the range of the perturbation. The extreme point of all connected phenomena thus advances beyond the exciting cause. In the extreme of the oscillation of the surface of the earth, its opposite expansions and contractions stretch beyond the normal hour. In the triple manifestation of the flow of force by the barometer, needle, and tension of vapor, the highest and the lowest range are not at noon and midnight, morning and evening, but at later hours. There can be in the economy of nature no sudden suspension of the motion of large masses. The transfer of force which accelerates or retards is a gradual process. No sphere could withstand a sudden change in the force which moves it; and we love to bring to mind the ease and elasticity of the motion of worlds, free to swerve in their own diameters and to bend in their orbits, yet every change having the safeguard of limit, the compensation of mutual adjustments. We see in the very elliptical form of the orbits of the planets the means of insuring this ease and freedom of motion. They are not bound in rigid circles, in which, by the fluctuations and transfers of force, the time of rotation and revolution would be increased or diminished, and the harmony of their movements be broken up. The application or withdrawal of force would, as it were, break a circle, and the planet would suddenly fall to or from the centre; while the ellipse is like an elastic spring. The elliptical orbit can approach nearer to or recede farther from the circle, and without a change of the major axis the motion can be quickened or retarded in different parts of the orbit, yet the period of the circuit be forever the same.

But to return: that there are two tidal elevations and two tidal depressions of the ocean waters, is known from the diurnal ebb and flow on the coast. How are these waves lifted up? In what manner are these alternate elevations and depressions of the great deep produced?

We will suppose that there is not an alternate rise and fall, but that the elevation is permanent, and is propagated round the world. This hypothesis is inadmissible; for the uplifted waters would rush along with a speed equal to the diurnal motion of the earth, and would overwhelm the land. We will suppose that the elevation is not permanent, but caused by a horizontal flow of water, — by currents heaping up the tidal wave. This is liable to the same objection, and such currents could not be formed without exhibiting their intense power. Only one other supposition can be made, — the water is lifted vertically, gradually raising itself up from its bed and then again subsiding to rest. This also is inadmissible; for, were the waters thus raised without horizontal currents, there must needs be a vacuum between the waters and the bottom of the sea. We cannot believe that to form the tidal wave the waters flow horizontally round the earth, or that they rise vertically from their resting place.

That which elevates the waters also elevates the earth on which they rest. It is the enlargement of the globe at equidistant points, and its contraction at the intermediate points, which rounds upward and curves downward the surface of the ocean. This is the reason why no elevation is noticed in inland seas and lakes, and why the *tidal wave* never manifests itself on the sea-coast, the perceived phenomena being simply the consequent ebb and flow of the surface water; for, as sea and land rise together, there is no gauge to mark the expansion and contraction of the globe. Nor does the curved surface of the ocean away from the land give the surface flow. Like minor waves rising by the possession of the due degree of force for the level to which they ascend, the waters are at the true level corresponding to their present force. The comparatively non-conducting land does not give out the required force, and near the land the unsupported waters establish the surface flow, and the more land-locked the waters of the ocean, as in bays and channels, the greater is this flow. In the Bay of Fundy the rise of the isolated waters is more than sixty feet, while the tide is hardly perceptible on the shores of the broad Pacific. The condition of oscillation is that the fall shall furnish the force for the rise. In the great vibration of the waters, the falling must transfer to the rising water its needed support; if, therefore, the connection be in part cut off,—if there intervene a barrier,—the force for support at the higher level may not be furnished. From this may perhaps come the horizontal flow. We present the idea for consideration, as one that may account for the known facts.

The tidal wave must rise at every half rotation of the earth, and the tidal depression must take place at the intermediate equi-distant hour. The time cannot change; this oscillation, being determined by the rotation of the globe, must coincide with the time of rotation. But the tidal flow, though caused by this alternation, is an independent oscillation. It is caused by it; but its movement is in different time and of a different nature. One is a horizontal flow of water for many miles, — the other, a vertical rise measured by feet. The one lags behind the other. They may be compared to two unequal pendulums, the longer impelled into vibration by the motion of the shorter, and their lengths bearing such a proportion that after a fixed number of beats, the motion becomes at regularly returning periods coincident in time.

To present a theory of the tides with minute exactness of detail, would require the labor of years exclusively devoted to the subject. We give the thoughts that have presented themselves to our mind, without an elaborate analysis of all the connected parts. We feel convinced that we have pointed in the direction of their cause, and that the connection of such a vast array of phenomena with the transfers of force, caused by the unequal orbits of the rotation and revolution of the several parts of the earth, proves that there is the spare force of rotation and revolution in constant diffusion, and that the earth, neither within her circumference nor in her orbit, is held in the rigid bonds of gravitation, but that an incessant and periodical oscillation is the life of the universe.

Before we leave this chapter, which we have devoted to the subject of the tides, we would once again recur to the opinion we expressed, that there are in all oscillatory movements two conditions of the undulation, — one, the impulse or cause; the other, the equipoise or balance, to receive the impulse. Thus, in the oscillations of the great tidal waves, we have for the impulse the transferable force of the earth's revolution in her orbit, passing from point to point in self-adjusting distribution, according as successively one hemisphere may require more, and the other hemisphere less, while the act of oscillation is the reception and transfer by the crest and valley of the wave of the rotary force of the world; the one receiving the more as it rises, the other imparting the more as it falls. Here, then, is a meeting point of these two forces of rotation and revolution; here they join in one movement; here is the place of intermingling, of intercommunication, so that the excess of the one can supply the lack of the other, and a perfect equipoise be established for the preservation of the harmony of these great movements.

Thus is presented to the mind, instead of two isolated independent forces, the one for revolution, the other for rotation, one majestic power, now divided according to the need of motion, now commingling in some common movement. Thus harmonious is nature, thus perfect in action is the creating thought of God, — a complete whole, in which even excess and defect, if they can occur, most perfectly balance each other.

We have no sympathy with theories by which the world is whirled about "with the dust of its own grinding," or its energy is impaired by the friction of its own action. We read such passages as the one we now quote with feelings of revulsion, though it is not unusual language: "Though we cannot by any stretch of the imagination embrace the period of the duration of the system, the conclusion still forces itself on the mind that at last it must end, that one by one the planets," by the action of gravitation, "must be lost in the sun, the sun itself perhaps be merged in other suns, to contribute by the immensity of its attraction to the destruction of other systems; that such will be the inevitable effects is admitted; but no inhabitant will witness the catastrophe, for long before that event our race will have been destroyed by the heat consequent on the reduction of the orbit of the earth."

Distrust all theories which involve the gradual construction, or the gradual wearing out of the universe. It came when God willed, perfect at the very first; it will continue without sign of weakness or mark of decay so long as He may determine, unworn and unimpaired, until He decrees ts end, when it shall have fulfilled His august design. Change, apparent to us, is the life-giving stamp of perpetuity.

> "The world's unwithered countenance
> Is bright as at creation's day."

There is a wide-spread belief that one of the planets of the solar system has been broken, — that the Asteroids are the fragments of that shattered globe. This idea meets us everywhere. It is expressed in standard books on astronomy, and in the rudimental treatises designed for district schools. What are the facts? How did this idea originate? Long since, in the days of Kepler, it was observed that the space between Mars and Jupiter was too extensive to be untenanted. There was a break in the series of planets, — the proportional distance was not maintained. One planet was wanted to give uniformity to the system, — to perfect the theory of relative distances.

On the discovery of the first Asteroids, the facts could be accommodated to the asserted law of distances, by the supposition that these small planets were parts of the one planet which *should* have occupied the space between Mars and Jupiter, — that by some internal convulsion this planet was rent asunder, or broken into pieces by a blow from some erratic comet. This idea of a God-deserted world left to the conflict of its own unrestrained forces, or wrongly placed so as to be in the way of some other sphere, was entertained for no other reason than that the supposition conformed the system to Bode's law of planetary distances.

We would not venture to assert that worlds may not, by omnipotent power, be recreated from the fragments of other worlds, — that the component parts of one sphere may not be remodelled into other spheres. But we cannot, merely to make good a mathematical calculation, entertain the idea of a broken planet with the angular fragments of its crust rounded into spheres by their own motion, revolving with orbits arranged and places assigned by the action of their centripetal and centrifugal forces. We need higher evidence than Bode's law. We regard the asteroidal system, the twelve or sixteen worlds revolving in their eccentric and beautifully interlaced orbits, as another striking and wonderful exhibition of the power and wisdom of Him who formed them, — this variety in unity, as another evidence of design, — this part of the heavens, as declaring afresh the glory of God, and showing forth His handiwork.

CHAPTER X.

"IN ORDER TO BIND TOGETHER FACTS, THEORY IS REQUISITE AS WELL AS OBSERVATION, — THE CORD AS WELL AS THE FAGOTS."
Whewell.

THE trade-winds, — which are supposed by some to be an aerial tide, caused by the attraction of the sun and moon, — oceanic currents, and some connected phenomena, now claim our attention.

In a former chapter we expressed the belief that there is a general law which governs the secondary movement of the planets as well as their primary movement; that is, that there is a normal intensity of rotary motion, as well as of the motion of revolution, belonging to every sphere, and that the one is in a fixed ratio to the other. But as, from its form, the intensity of rotation decreases from the surface of a globe to its axis of motion, the law which governs this movement must be related to the mean velocity, and not to the velocity of either extreme, — to the average speed, not to the motions of either the equatorial or polar regions.

From this consideration, it is evident that while, in the earth, for instance, there exists the due amount of rotary force, this force is unequally distributed through its mass. There are therefore portions which have a greater, and those which have a less degree of force, than the mean force belonging to the sphere. From the tendency of force to diffuse itself equally through every mass of matter, and from the law that force is transferred from matter not susceptible of motion by it to matter free to move, it follows, as force is distributed through the earth according to the capacity for motion, that air and water, to the extent in which they are free to move independently of the more solid parts of the earth, will have at the poles a motion greater than that of the earth, and that these elements at the equator will have less motion than the earth.

To illustrate these ideas: Were the earth a perfect sphere, the mean surface velocity of rotation would be found at the parallels of 45° north and south latitude, and, as the earth is shaped, this determination is sufficiently accurate for the purpose of general illustration. From this line of mean velocity of surface rotation, towards the poles as well as towards the equator, the rotary force exists in varying degrees, according to the capacity for motion of the different parts rotating as a united sphere. But the force will not be thus distributed to that matter, which at the equatorial or polar regions is not necessarily a rigorously connected part of the earth, and which may have some capacity for motion independent of the motion of the sphere. To the degree, therefore, in which air and water have the power of independent motion, this motion will be found to be related in its action to the mean surface velocity.

The atmosphere has this freedom of independent motion. In its movement, therefore, it will be found that the mean velocity of rotation, — that with which the surface of the earth moves at 45° latitude, — will be the governing principle of its independent motion. It will limit its eccentricity. The winds, of course modified by other laws, will, as a general principle, be related to this mean of rotation, and a reference to this fact will be found to dissipate some of the mysteries of the movements of this free element. The atmosphere at the north polar regions and the adjoining parts of the temperate zone may have a force of motion more intense than the the rotary force of those portions of the earth. It will therefore move southward to increase the orbit of rotation, and eastward because it moves faster than the earth. Hence the prevalence of northwest winds in those regions. As we proceed southward through the mean of rotation, there is found a great belt of variable winds. "What," says Mr. Guyot, in reference to these regions, "is more fickle and capricious than the winds? They are symbols of changeableness itself." At the equatorial regions, the atmosphere has less force of rotation than the surface of the earth on which it rests; of course it has an apparent motion westward, from moving more slowly than the earth. This wind also blows from the centre northward and southward, thereby narrowing its orbit of rotation, — the converse of the action of the winds from the poles, which move southward and northward to widen the orbit of rotation. The easterly trade winds, by the clouds and vapor which they transport, may present the aspect of two zones, which seen from the moon might appear like the belts of Jupiter, which planet, from its immense diameter, and the consequently great velocity of surface rotation, would most distinctly present the zones of retarded and accelerated atmospheric motion.

We notice in the Annual of Science that Mr. Babinet refers to retarded and accelerated rotation as the cause of oceanic currents, believing "that the waters which flow from the equator have an inclination to advance before the motion of the earth, whilst those which come from the poles have a tendency to remain behind the motion." But we apprehend that the reverse is the case. The waters follow the laws of the atmosphere, and in their respective motions there is that resemblance which indicates the same general law. There is the great westerly current of the equatorial regions corresponding to the trade winds. This extensive current is deflected northward or southward, when it strikes the eastern shores of the continents, and thus its waters narrow their orbit of rotation by advancing towards the poles. The currents flowing from the polar regions widen their orbit by their advance towards the equator. At the point of equilibrium of rotary force, about the parallels of 45° north and south, all the oceanic currents are turned to the eastward. Thus, in this general eastward movement is found the compensation for the great westerly current of the tropics, and the equilibrium of the waters is preserved.

It has been supposed that the Gulf Stream, which is a branch of the great westerly current arrested by land and turned northward along the coast of the United States, is deflected to the eastward in latitude 45° by the Banks, which extends across the Atlantic Ocean in about this parallel. We are inclined to the opinion that the same cause which turns the current also forms the Banks.* This deflection is not local and accidental. The current flowing by the eastern shores of Asia bends to the eastward at the same par-

* It is said that soundings extend from the Newfoundland Banks across the Atlantic Ocean. We are also informed by an experienced navigator, that in the summer season the *Gulf Stream flows* farther north before its bend to the eastward than it does in the winter, thus following the sun as do the trade-winds, — these phenomena being related in their changes to the changing angle of rotation and revolution, the extreme points of the earth in their revolution following the direction of the sun as the centre of revolution.

allel; in the same degree of south latitude, a polar current flowing northward bends to the eastward off the Cape of Good Hope, while inshore close to the Cape, there is a westerly current, these opposite currents running side by side. The parallels of 45° are the dividing lines; currents in lower latitudes flow westerly, in higher latitudes flow north and south with an easterly tendency, until they reach this point when they are turned eastward, the waters on one side of the equilibrium of force having a greater, and on the other side a less intenseness of velocity, than the solid earth. The deviations from this general law are caused by the position of the coasts against which the waters impinge. In the map of currents presented in Guyot's "Earth and Man," every arrow showing their direction conforms to this statement. The moving elements, air and water, in the polar regions move with greater, in the equatorial regions with less, than the mean force of the rotation of the earth. Why, then, trace the cause of the trade winds to the rarefaction of the atmosphere, caused by the more intense action of the sun's rays within the tropics? Why speak of the attraction of the sun and moon as causing aerial tides, or oceanic currents, or tidal waves? The cause of these phenomena will be found on the earth, not in the skies, — in the changing force of the motion of our own planet, not in the attractive power of other spheres.

Geology may find, in a reference to the parallels of mean rotary velocity, some help in the investigation of the structural changes of the earth. The erratic drift of boulders and diluvium is spread over the frigid zone and over the northerly parts of the temperate zone of the northern hemisphere. This dispersion of drift, we believe, was simultaneous. The torrent rushed from the polar region southward, with an inclination eastward. Its southerly bounds are about the parallel of 45°. Again; the range of the drift of icebergs seems to be determined by their equilibrium parallels. The mean temperature of the earth at the poles is said to be about 10° of Fahrenheit, and about two or three degrees below zero at the two poles of maximum cold, twelve degrees distant from the poles of the earth's axis. The Arctic and Antarctic Oceans are frozen to a very great depth, and "during eight months of the year a continuous body of ice extends round the poles of maximum cold, occupying a sort of elliptical area of above four thousand miles of diameter." Icebergs are detached from the margin of this frozen zone at certain seasons. By a beautiful provision of nature, ice cooled to a certain degree changes its character, and from being a non-conductor becomes a conductor of force; at one time, as if it were a mantle spread over the earth and sea, it confines the heat until the extreme is beyond the limit which nature has assigned; it then is no longer a barrier to the required adjustment,—and perhaps by this very means, some transfer of force rends the masses of icebergs from the parent bed of ice, and urged by the present force they move slowly but strongly towards the south or north, undisturbed in their advance by storm or current, until by the widened orbit of rotation the force which impelled them from the poles is absorbed. Icebergs have been observed thirteen miles in length, with perpendicular sides rising at least one hundred feet above the surface of the sea; but they are found of all sizes and shapes drifting from the frozen oceans. This drift of ice is arrested mainly at about the parallels of 45°, though solitary masses may occasionally stray farther from the poles. The icebergs of the northern

Atlantic, it is well known, collect at the Newfoundland Banks as their resting place; for there the mean of rotation is obtained.

We notice the same limit to the drift on the land, — to the transportation of diluvium and boulders from the north. The course of this mineral or earthy flow was southeasterly. Its path is written down in grooves and striæ, or diluvial scratches, and by the smoothing of the northern sides of the mountains. This current also arrested its flow when the impelling force was reabsorbed in rotation by the extension of its diurnal orbit. But the flow of earth and stone was not like the slow movements of the icebergs; by a sudden rush, as it were, the mountain masses fell from the north. The one is from the every-day action of force, — the other came from a readjustment of the earth's surface when it took the form which fitted it for the home of man. The limit of the drift is, however, the mean parallel of latitude, — the boulders and diluvium in northern Europe and northern America being confined to the Arctic zone, and to the northern part of the temperate zone. There is no easterly or westerly drift, and none beyond the parallels of mean latitude, or none of great extent arising from any widely acting cause.

The diluvial scratches or striæ marking the track of the erratics from the north, are straight as a mathematical line. They are found on the plains, on the northerly side, and occasionally on the south side of mountains. By whatever force they were moved, their course was not obstructed by the inequalities over which they passed. They flowed on as if without weight, — now lightly, now heavily touching the surface of the earth, ascending or descending as more or less upborne by the impelling force.

We believe that these mountain masses moved with intense velocity, because the striæ were so evenly grooved. No current of water, nor iceberg floating in the current, nor slide or glacier, could, from known causes move with sufficient velocity to form these grooves. There would be irregularities of motion, depressions, gyrations. There is another proof of the intense velocity of the motion of this flood of rocks. The edges of the engraving rock, the angular points which made the record, did not wear down or become blunted. The only wear was on the tablet which received the impression. The effect of great velocity is to preserve the cohesion of the moving body or instrument. A tallow candle, soft as it is, when discharged from a gun will pierce through a pine board. Thus, by intense velocity alone could the sharp angle of the rock making the impression be preserved. A slow motion of the mass would have worn the engraving, as well as worn into the engraved rock, and there could not have been an equal and even groove made for any considerable distance. It has been remarked, that the striæ might have been caused by small stones, pebbles, or sand, between the drifting boulders and the rocks over which they moved. But such intervening substances would have been crushed by the weight and friction. Smoothing or polishing to some extent might have been produced by this means, but not the straight furrow.

The *vortex moulds* (pot holes) to which we have before referred, are circular cavities formed in the solid rock. These generally contain a rounded stone or stones. On the shores of Lake Superior are many of these cavities, and we are told that "a stone in one of these might have weighed fifty pounds. Some of the holes were three or four feet deep, and as many in diameter; some of them formed steps, the stone having worn down at one side of the hole, and worked on horizontally awhile, and then downwards again."

It has been surmised that these holes were worn into the rock by the action of water on stones lodged in its crevices. How could waves or running water make stationary whirlpools sufficient to whirl stones weighing fifty pounds penned up in a narrow crevice? Yet only a swift rotary motion of the encaged stones could have worn these circular holes.

The motion required to produce these smoothly rounded excavations could hardly be induced by the action of water. The stones which formed them were probably arrested fragments of the rushing flood of rocks. Their onward motion being suspended, they would revolve with the same force which before impelled them in a right line. They would act as the arrested water of a current which exhausts its force in the whirlpool; and the existence of such excavations seems to prove the intense velocity of the erratic drift.

We refer the period of the drift to the time of the great change in the earth, which under the Providence of God gave to it its present form. The moving mountains acted upon by a force more than sufficient for rotation at their position rushed southward, with an easterly inclination widening their orbit, and moving with great velocity until they arrived at the place of rest.

If this hypothesis be unsound, we believe that no theory, which does not refer the rush to a current of the most intense velocity, can explain the erratic drift. The respect we entertain for those who seek to account for the facts of the case from natural every-day causes, leads to hesitation, and an analogous action on a smaller scale would confirm the doubt; but the record on the mountain-sides, the characters uneffaced by time impressed on the granite tablet, tell us of a power no longer in operation, the action of which was never observed by man.

Meteorology, if it may be allowed the name of science, is a science of facts only. The meteorologist neither foretells the weather, nor explains its phenomena. He sits, patiently musing over storm and calm, tornado and zephyr, fog and sunshine, the arctic cold and the torrid heat, looking anon at the weathercock and the barometer, diligent in recording variations, in a note-book already full of facts, without a principle for their arrangement or explanation,—waiting for light to shine forth from his recorded observations upon the vicissitudes of the weather.

Yet there is nothing out of rule, beyond law, in these apparently capricious changes. There is nothing in this world of ours without a cause. The vane never turns but by the operation of law altering the direction of the wind; nor are the laws which govern the gale and the storm so complicated and mysterious that there is no hope that the mind may comprehend them. In olden times the heavenly bodies presented to the observer nothing but involved, confused, entangled movements; yet the harmonious motion of the spheres now ascertained gives to the astronomer his deepest impression of the simple majesty of nature. If, a century ago, it had been asserted that the time would come when an almost instantaneous communication of thought could be made to a friend a thousand miles off, and that ideas would rush across continents with the speed of lightning, the expression of this now established fact would have been deemed as wild and chimerical as it may seem at present to say that future ages will so comprehend the laws of the elements, that to a very great and practically important extent, the common weather will be predicted. Compare any one science as it now is with its condition centuries ago, and we hardly dare to limit its possible advance; for its progress is with accumulative velocity; discovery does not exhaust the field; every instance of successful inquiry gives a vantage ground for greater results. Why should we doubt? "Even Æolus reigns by law," and there is a control over the wildest elements. The law which directs and restrains has not been promulgated, yet we witness the exercise of this control. In every truce which terminates the struggle of the elements we recognize, though we may not understand, the agency which effects it.

We quote the following from an article in the North British Review: "For weeks and months the hourly variations [of heat] are of the most capricious and irregular character; the thermometer being sometimes stationary for a day; sometimes highest at midnight, and lowest at noon. When we combine however the three hundred and sixty-five observations at each hour, we find these hourly means are the ordinates of a curve of beautiful regularity, each of the four branches of which do not differ more than a quarter of a degree of Fahrenheit from parabolas. The varying temperature of a day, therefore, with all its starts and irregularities, is controlled by a law as precise and unerring as that which regulates the planetary motions. But the other agencies of the atmosphere have not been subjected to law. The wind and the rain, the hail and the snow, the storm and the tempest, seem to have a will of their own, and to triumph in mockery over the person and property of man; yet lawless as they appear, we shall some time or other discover their haunts and disclose their secrets, even though we may never succeed in disarming their fury and reducing their power. Even the hurricane and tornado must yet submit to intellectual control, nor shall the lightnings of heaven escape the analysis of genius. . . . The time will come when the electric flash will be predicted with as much certainty as the occultation of a star, or the emersion of a satellite from the shadow of its planet."

If in any department of inquiry there be not an even advance, if any science lags behind the others, there is great reason to suspect that there is some radical defect in the foundation-theories on which that science is based; its cultivators should therefore go back and reëxamine the first principles on which it rests; these readjusted with accuracy, they may be able to hold even way with their associated laborers in the much divided field of inquiry.

Upon a reëxamination of first principles, it may be found that the defect to be remedied in the department of meteorology is a mistaken idea of the extent of the influence of the sun, both in relation to his attraction and to the heat which emanates from him. It may be that the meteorologist goes too far from the field of action, for the cause of action, — that he traces too much to foreign influence, too little to domestic changes.

We quote from Sir John Herschell a description of the supposed immense range of the sun's influence. "The sun's rays are the ultimate source of almost every motion which takes place on the surface of the globe. By its heat are produced all winds, and the disturbances which result in the changes of the electric condition of the atmosphere. By these rays the waters of the sea are made to circulate in vapor. By them are produced all the disturbances of the chemical equilibrium." And the paragraph concludes by assigning the phenomena of volcanic action, indirectly, to the rays of the sun. Even more sweepingly are all the changes of weather made dependent on solar influence. We can hardly call to mind a single fact in this department that is not explained by the vacua of the atmosphere, — by the unequal rarefaction of the air, produced by varying temperatures.

The winds have but this one general cause assigned to them; and yet, if there are as great inequalities of density in the atmosphere as are supposed to exist, it does not clearly appear how these could create the winds, with their intensity of force, long duration, and extensive range of motion. An effect must be, in some degree at least, commensurate with its cause. Motion cannot produce or prolong itself without cause. To fill vacua, — to supply additional air for the restoration of equilibrium to the rarefied parts of the atmosphere, would not require the rush of the colder air for months, for weeks, for days, or even for hours. The repairing and equalizing process, as the winds now blow, would cause wider breaches of continuity than are repaired, — would create more void places than are filled. This would undoubtedly be the case when the horizontal currents of wind are greater than the upward-flowing current of heated air, which is supposed to cause the wind. Vacua are filled, and the equilibrium of the density of the atmosphere is preserved, by the elasticity of the air. This acts, as it were at once, almost instantaneously; for the action of elasticity is as rapid as the propagation of sound, while gradually heated air ascends slowly, its place being supplied as slowly. It can never create the intense motion of a wind.

The colder air does not rush toward the heated air.

Did it so rush, there would be established a violent wind, regular and unintermitting, from the poles to the equator; there would be no wind, as there often is, speeding from the sunny plains of the south, over snow-clad mountains, toward the region of perpetual ice; the trade winds would not gently flow, spreading themselves toward the colder north and south; nor would the winds of the frigid zones incline to the eastward. It is true, that there is a draft of air to supply combustion, and that, whenever heated air ascends, other air will move to maintain the equilibrium. The comparative density of the air, however, cannot account for the geographical or periodic winds, or for local occasional storms. It cannot explain either the direction, intensity, or duration of the wind. Take, for an instance, the circular motion of a tornado; how can this movement be induced by a vacuum, or how can it restore the equilibrium of the atmosphere?

The idea has been recently entertained by scientific men, that the trade winds arise from retarded rotation. It is evident that the same cause inclines them northward or southward, as, by thus narrowing the orbit of rotation, the equilibrium of motion is restored, since the degree of present force determines the orbit. Thus the winds from the frigid zones flow to the east, moving with greater velocity than the earth, and southwardly or northwardly to widen the circuit. Thus the degree of force possessed by air compared with the velocity of rotation of the part of the earth over which it moves, is not the cause of the east and west winds only, but it also determines the direction of the north and south winds. The inequality of force, which thus gives direction and intensity to the permanent geographical winds, can be traced directly to the differing degrees of rotary velocity of different parts of the globe; and we hope to be successful in tracing the inequality which gives rise to the variable and occasional winds.

The land and sea breezes alternating with day and night, are frequently cited as proof that the varying density of the air, caused by differences of temperature, determines the direction of the wind. We think that this alternation is an illustration of the views which we present. By night a greater force of revolution is required, by day a less, as the part of the earth is turned from or towards the centre of revolution. The regular land and sea breezes are most uniform and most active between the tropics, where day and night make the greatest difference in the orbit of revolution. These changes of the orbit are accompanied by a transfer of force, and, of course, from the different conducting powers of land and water land and sea breezes would ensue.

The monsoons are periodic winds, which in some regions blow regularly for a part of the year from the sea landward, and at the opposite season blow from the land toward the sea. In the regions where they occur, for a part of the year the meridian sun is vertical over the land, and for a part of the year it is vertical over the water. This is the condition of these periodical alternations. The changing position of the sun indicates that the earth's axis of rotation has changed its position in relation to the centre of her orbit. Of course the angle which rotation makes with revolution has also changed, and, where monsoons occur, the extremes of revolution periodically pass from the ocean to the land, and from the land to the ocean. The change of the position of the sun, north and south of the equator, causes, or rather, denotes the change of the flow of the spare force of revolution which moves the elements. This determines the monsoons. The trade winds also, in some degree, follow the course of the sun; in our latitudes the northwest winds are most frequent and violent in the winter season; the Gulf Stream in summer runs further north before it bends to the eastward. All the phenomena thus "following the sun," and supposed to result from the influence of his rays, and the consequent inequalities of temperature, can thus be traced to the changes of relative motion of different portions of the globe, which alter the direction of the currents of force, and consequently of the moving elements, air and water. The position of the sun indicates these changes, but neither his comparative distance, nor the comparative directness or obliquity of his rays produces them. As the hands of a clock indicate his position, so does his position indicate the changes of the motion of the earth, and the changing orbits of revolution and rotation of the several parts of the globe. From these changes, and from the consequent changes of the flow of force, come the movements of the elements which give us the variations of meteorological phenomena.

We are fully aware that we have advanced many principles, which contradict the opinions of those who have devoted themselves to the examination of these subjects. In meteorological science, however, we do not feel that we are in the opposition. It is confessedly an open field of investigation. There is little or nothing definitely settled, and, if new theories afford any clew to explanation, this result may strengthen these theories as applied to subjects, which are generally believed to be well explained by the theory of universal gravitation.

A question has been recently proposed from a high

source, "whether the gush of rain which usually follows the detonation of a thunder storm is the effect or the cause of the electric flash." Should there not be principles competent to answer such inquiries, or rather principles, so thoroughly settled, so accurately determined, as forever to prevent such questions from being asked by scientific men? This is, however, but one of the many points on which nothing is distinctly known. What creates the progressive increase of heat from the arctic to the torrid zones? Can the comparative obliquity of the sun's rays give a full and satisfactory explanation? Is there not a source of heat in the earth, independent of the rays of the sun? What is the cause of the varying degrees of humidity in the atmosphere? Why does water change to vapor more rapidly at one time than at another? Why do the vapor wreaths now hang over the surface of the earth, and now float aloft as clouds? Whence the cirrus, half-illuminated clouds? Why are the magnetic poles the poles of maximum cold? Why do the curves of equal temperature have for their poles the poles of revolution? All these phenomena are surely not caused by the inequality of the sun's rays, their varying obliquity, and the consequently irregular density of the atmosphere.

In the last chapter we alluded to the many provisions, creating the transfer of force, from which proceed the life-giving alternations of the world. One of these provisions was the diameter considered in its proportion to the orbit of revolution, and another, the velocity of rotation. In our planet, the greatest diameter, — giving the widest extreme of the differing orbit of revolution of its opposite parts, and the most intense rotation, — is between the tropics, — at the equator. In that region consequently will occur the greatest and most rapid transfers of force. Accordingly, we find this indicated, not only by the needle and barometer shewing even the hourly variations of the strength of the flow, but by the greatest and most sudden alternations of all meteorological phenomena. There are the most active volcanoes, the most frequent convulsions of the earth and disturbances of the electrical condition of the atmosphere, and the most violent hurricanes. For one instance, at certain seasons, in some places between the tropics, almost every day, some four hours after the sun has attained the meridian, there is a thunder storm, while so rare is this occurrence in the northern regions, that a traveller mentions as a fact worthy of note, that in Nova Zembla and Spitzbergen "it is sometimes heard to thunder." Contrast the general meteorological phenomena of the polar regions with those of the torrid zone. They present the extremes of a habitable earth, and these extremes, we believe, are caused by the inequality in the degree of those transfers of force, which give both the periodic and the occasional alternations of the elements; and, in tracing out the causes of meteorological phenomena, we are led to look to these transfers of force. For instance, if we would know what causes the gush of rain accompanying the flash of lightning, we should consider that the cloud is formed from water by the reception of force, that by its rapid condensation force is given out suddenly in the flash of lightning, and that the condensed water falls at one and the same time; — they are the different phases of the same act.

We regard the barometer, not as indicating the pressure of a column of air over it, but as indicating the level of its rotation, and we believe that its oscillations at the same apparent altitude mark the rising and falling surface of the earth. By this view we have an explanation of the connection of the changes of the height of the mercury with the changes of the weather.

The rise and fall of the mercury denote the different levels at which every part of the surface of the earth moves in its diurnal round. By the rise of any portion of the earth's crust more force for motion is required, by its fall less is needed. In the one case force is given out or transferred from the earth; in the other it is absorbed. A high range of the barometer indicates a low surface and dry weather; a low range shows a high surface and wet weather. A high surface therefore withdraws force from the vapor of the air, and it condenses; from the low surface force escapes from the water in the form of vapor. To repeat the idea, a high surface of rotation, indicated by the low barometer, withdraws the sustaining force of vapor; a low surface, on the other hand, indicated by the high barometer, supplies the force which gives tension to the vapor. The act of withdrawing force by a rise of the earth would recondense vapor; the act of supplying force by a fall of the earth would sustain the vapor in high-floating clouds. Thus is explained the varying height at which the clouds float, now enveloping the surface with fog and mist, and now rising miles above us, with sharp and well defined outlines.

Water exists in the atmosphere in two states; in the one, at its greatest tension it is the unobserved humidity of the air; in the other, it is partially condensed in the form of mist, or fog, or more properly vapor. The humidity of the air is shown by the condensation that takes place in the driest weather, on the outside of a vessel containing water so much cooler than the atmosphere as to cause this condensation. The dew-point is that degree of difference of heat between the atmosphere and the earth, which is sufficient to cause the condensation of the humidity of the air into dew. The formation of vapor and its recondensation are not, however, the transfer of sensible heat only. They are caused by the transfer of this heat, of electricity, of galvanism, and of the magnetic fluid, all of which are but phases of the dynamic power of the world, — branches of the great current of force by which the equilibrium of motion is preserved. Thus dew is formed by night on those parts of the earth which, being turned away from the sun, require the greater force of revolution on account of their enlarged orbit. It is not formed in cloudy weather, because from local causes, as evinced by the very existence of the clouds, the force which would be given out from the formation of the dew is not required. Even the vapor which rises from a water-fall gives to the air of its vicinity an electrical state different from that of the adjacent atmosphere. The electrometer proves that the electrical state of the air changes with the condensation that causes the fall of rain. "In fogs, and in the commencement of a fall of snow," says Humboldt, "I have seen by a very long series of observations, the previously permanent positive electricity rapidly pass into the negative, both at the plains of the colder zones and at great elevations between the tropics." We know that the tension of vapor has its two maxima and minima daily, which correspond to the maxima and minima of the diurnal alternations of the surface of the earth.*

* We have frequently alluded to the condensation of vapor by mountains and high lands, for the reason that they require a greater force of rotation. The fact is thus clearly expressed by M. Guyot: "The mountain chains are, then, the great condensers placed by nature, here and there along the continents to rob the winds of their treasures, to act as reservoirs

From this cause we have the different character of the two opposite winds of our climate, the north-west, and the south-east. The one is the polar current, in which the air comes from the north, inclining to the east because of accelerated rotation; in the other the air comes from the south, with an inclination to the west from retarded rotation. The one has more, the other less, than the normal force; the one by the additional force increases the tension of vapor, the other by the diminished force decreases it; the one imparts force, and water passes into vapor, the other withdraws force and the vapor condenses; the one usually gives dry and clear weather, the other, fog, rain or snow. Of course the character of the wind is modified in some degree by its passage over water or land, but, notwithstanding this modification, the characteristic difference of these winds remains, and is proved, not only by the different tension of vapor, but by their contrary effects on the animal organization. The one is an exhilarating, exciting wind; the other is unelastic and depressing.*

for the rain waters, and to distribute them afterwards, as they are needed over the surrounding plains." . . . "But as we have said the plateaus have also a marked effect on the distribution of the rain waters. Their borders act as mountains." . . . .

"The effect of the condensation in the supply of the force of rotation is shown by the change it produces, in the character of the winds of the tropics. The trade wind is from retarded rotation of the atmosphere. The copious condensation of the moisture of the air into clouds, and of the clouds into rain, at the rainy season, breaks up the trade winds. When the trade wind blows with its accustomed regularity there is no condensation, 'the air is dry and the atmosphere cloudless.' The periodical rains of the tropics follow the sun, or rather follow the extreme diameter of the revolution of the earth."

* "This conflict of polar and equatorial winds," says M. Guyot, "opposite in character and direction, gives to our climate one of its most char-

How various are the considerations, how numerous are the facts, which seem to support the position, that vaporization and condensation are among the means of preserving the equilibrium of motion! The bright and clear sky, the dark and lowering cloud, dry weather, fog and mist, the parched earth, and the copious rains, — these ever-varying alternations speak to us not of the changing heat of the sun only, nor of a lighter or heavier atmosphere, but of the ever-passing currents of force, and through them of the varying level of rotation as indicated by the barometer. Who can recall the high electrical state of a clear day in midwinter or midsummer, the damp easterly winds of spring, or the murky atmosphere of the dog-days, and not perceive that the tension of vapor is determined in some degree by the varying motions of the earth, as she speeds in the different parts of her orbit?

We divide the winds into two classes, — the normal winds of the region where they occur, and the occasional winds arising from local disturbances of the level of rotation. At the extremes of rotary velocity, in the equatorial and polar regions, the prevailing winds are of the former class. There are the regular tropical winds, and the regular polar currents of the northern and southern regions, arising from the retarded or accelerated motion of the atmosphere. Midway there is no prevailing atmospheric current, and the winds arising from local changes appear fitful and capricious. The regular winds are far-pervading, and blow in right lines

acteristic features, its changeableness, its extreme variety of temperature, of dryness and moisture, of fair weather and foul." . . . . "The polar winds will prevail in a country and will endanger the crops by the prolonged dryness of the atmosphere, while further east or west, the trade winds will spread fertility by beneficent rains."

or in curves of wide radii; the occasional cover a more limited range, moving in narrow circles. The one class has the character of uniformity; those of the other are storm winds. Thus the hurricane of the tropics is usually a westerly wind; in our regions, and especially further north, the easterly winds bring the most frequent tempests. Storms being induced by sudden changes of level, the air moves with the newly acquired force in a circle, and, as the depression propagates its undulation, the circle advances. The cause preceding the effect, and the change of level being the cause, the barometer indicates the approaching tempest by its sudden oscillation. Storms accompanying the prevailing winds are occasioned by local disturbances accelerating these winds.

We regard, then, the changes in the direction and intensity of the winds, as determined by the transfers of force from altered levels of rotation, and by changes in the position of the earth. The theory which we present for consideration will not only thus explain the great phenomena of the weather, but also many of its apparently fitful changes. We have a reason, for instance, for the want of permanency of fair weather, when the wind of our climate *backs in*, as it is called, or comes round by the north to the westward after an easterly storm; for this takes place by a local change of level, not by the restoration of the normal wind as determined by the season and by geographical position. We have also the reason for the equinoctial storms, which result from a change of the position of the earth relatively to the sun, the centre of revolution, giving a new direction to the flow of force; and also for the changes of the weather connected with the phases of the moon. The moon is the people's barometer; and we believe that it not only widens our local system and increases the flow of force, but periodically increases and diminishes the intensity of the flow. Changes of weather result from the transfers of force, and these are determined by the motion of the world, as the moon and the earth swerve sympathetically in their common orbit of revolution.

We pass to the consideration of other phenomena, which seem to strengthen the position that the different degrees of the tension of vapor, and the direction and intensity of the winds, are determined by the degree of force relatively possessed by earth, water, and air, and that the tempest and storm are but transfers to preserve the equilibrium of motion, — that they are not accidents, but the consequence of those alternations which give to the elements their life and activity, — change within a fixed limit, creating the "agreeable fitness" of all things in this beautiful home of man.

There have been earthquakes in every part of the world, and they are not rigorously confined to any season of the year. They are, however, by far the most frequent in the torrid zone, and at the commencement of the rainy season, at the vernal and autumnal equinoxes, and at the changes of the monsoons. We trace them, then, chiefly in regions where the transfers of force are the greatest, and at the times of such changes in the earth's position as vary her motion, and give alternations to the flow of force. An earthquake sometimes changes the level of rotation of a distant part of the earth, in addition to the local convulsions which it produces. After a violent earthquake in the island of Sumatra, Darwin traced out, as a consequence of it, a subsidence of land at a point six hundred miles distant. There have been permanent elevations of a vast extent of coast by this means. Earthquakes are not always preceded by atmospheric changes, but are invariably succeeded by rains, even when rain is out of season where they occur. Very little is known of the cause of earthquakes; but we will quote a remark of Pliny, which is somewhat in accordance with our views. "They are because the elastic forces, concussive by their tension, accumulate in the earth when they are absent in the atmosphere;" in other words, the force of motion is too great for the orbit of rotation, and the earth rises suddenly and convulsively.

Volcanoes are attended with the same phenomena as earthquakes. In their periods of activity, beside the throwing up of great volumes of lava and other substances, and the local agitations produced, there is frequently an agitation of the earth, which is perceived often at a very great distance. The eruption of a volcano is attended "with fierce winds and deluging rains." Stromboli was considered as the dwelling-place of Æolus, the regulator of the winds; and the fishermen on the coast were able to foretell the weather by its eruptions, or by the premonitory symptoms of an eruption. The connection between the eruption of even a small volcano and the state of the barometer is now generally acknowledged. The earth in the vicinity changes its level before and during the eruption.

The connection between the undulations of the surface of the earth and the weather, thus indicated by the earthquake and the volcano is not an exception to a general law. It is, if we may be allowed the expression, the abnormal action of the law; it is a fact, standing as it were unnaturally, out of position, thereby distinctly showing the action of a law otherwise hidden under the uniformity of its usual routine. This known sympathy between the weather and the change of level of a part of the earth's surface is not occasional or accidental. By these convulsions sudden changes in the weather are produced; but it is also varied by more gentle alternations. It is by the transfers of force that the rain falls and the winds blow. It is by the health-giving readjustments of the force of motion that we have the occasional storm; it is by the equable flow of the life of the world that we have the usual serenity of the face of Nature.

The mean temperature of a season often changes over a large portion of the earth. We have perhaps a cold summer, or a comparatively mild winter. This inequality spreads as from a centre. As we go from this centre, the excess or deficiency of heat decreases; and, when we pass the limit, the increase or diminution changes conversely, and we have the opposite increasing alteration of the mean temperature. This undulation of the curve of mean temperature moves mainly from north to south. How is this undulation to be accounted for? The sun shines as in former seasons, nor can atmospheric pressure be tortured into a reason.

We present another phenomenon connected with the transmission of force, for a hasty examination. The aurora borealis of our northern regions appears to be an arch or circuit of the "magnetic force," spanning a portion of the earth, as if conducting the flow from whence it abounds to where it is needed without passing through the intermediate parts. The shooting cylinders of rays have been compared to the flame which arises in the closed circuit of the voltaic pile between two points of carbon. Is it not a flame stretching between the surcharged and undercharged parts of the earth or atmosphere? That it is a passage of what is called the galvanic fluid, we know; for it disturbs the magnetic needle, which often points to the corona of the arch. It indicates a disturbed electric condition ; and, while the lightning in the electrical storm shows a sudden transfer of force, the aurora represents a more gradual adjustment of the equilibrium. It does not occur in the night-time only, but its path may often be traced by day in a circle of half-illuminated clouds. It is usually accompanied by a high barometer and great tension of vapor, — by all the phases of motion which indicate more than the usually present force. For this reason we consider it as a distribution of the magnetism of the polar regions, passing over an already well supplied region. With us the aurora most frequently appears after an easterly wind has subsided, and a clear wind from the northwest has begun to diffuse the spare force of the more northerly regions.

The transfer of force shows itself often in a light which follows its path, and attends its movements. There have been luminous fogs; there are often clouds half-illuminated, shining not by reflected light, but by their own emanations. Without any other assignable cause, whole nights, as in Italy and the north of Germany in the year 1831, have been so light that the smallest print could be read at midnight. What shall we say of the phosphorescence of the sea, the waters of which, disturbed by the prow of a ship, with a fresh wind and clear sky, are like waves of light? The presence of animalculæ will not account for this; for a certain electrical state of the atmosphere always attends it.

But we must leave this, to us, interesting branch of our inquiry. We know that we must have often erred in the application of the principles which we present; but errors of application should not detract from the value of a general principle. Our hope is, that in our illustrations we may have been so far successful as to give the necessary explanations of our views, so that they may be examined and tested, and, if found worthy, be applied by those who, from years of devotion to scientific pursuits, are more competent to work. Atmospheric changes seem capricious, as if without the law; but this must be in appearance only. The facts which nature presents are never out of rule. As science advances, the discordant and conflicting phenomena decrease in number; the periodical, unvarying, regulated, become more numerous; order extends itself, and accident gradually puts on the appearance of design. It appears as if order was created by the sympathy of the mind with nature; but now and ever, whether we perceive it or not, a sovereign law is completely and perfectly obeyed. The change is in ourselves; and the perfection of philosophy is this transfer of the order without to order within, the regularity and harmony of nature giving distinctness and definiteness to thought.

Before we close this chapter, we will give the reasons why we disbelieve altogether in the existence of the supposed intense heat of the interior of our globe. This heat would be a useless force, and, so far as the action of nature can be traced out, there is found no unemployed or reserved dynamic power. All force is in unremitting action. There is no waste energy, none pent up, confined, laid by for ulterior use. The sensible or diffusible heat of the planet appears to come from the vital force constantly circulating through every part of it, and through every element which rests upon it, or envelops it. A flow of heat from the sun, the centre of its revolution, warms its surface, and beneath its crust the magnetism of the earth, from the decrease of rotary force, gradually becomes sensible heat as we descend. The magnetic elements of any portion of the globe may be calculated from the heat of that portion, and conversely the distribution of the magnetic force may be calculated from the distribution of heat. Force exists everywhere as it is required, and resolves itself into sensible heat no further than by the transfers needed for readjustments of equilibrium. The normal action of the needle is from the steady flow of force, and its variations of intensity are from the periodic or occasional variations in the current. On the other hand, the supposition of intense internal fire requires the belief of a permanent inequality, — an abiding want of equipoise.*

We shall be pointed to the earthquake and asked, whence its power? We reply, it is a sudden restoration of equilibrium, proving that the globe does not bear even slight irregularities, but frees itself of all superfluous power. In relation to the volcano, we have the same answer; and further, the force or heat exhibited by the earthquake and volcano is altogether insufficient, taking into consideration the magnitude of the globe, to indicate that its interior is a mass of liquid fire.

We shall be pointed to the crystalline structure of the Plutonic rocks, as proof that these rocks were once fluid from intense heat, and to the upturned strata of the globe, to its distorted mountain ridges, as evidence that heat was

* The writer of a valuable scientific article in the fifth volume of the North British Review, expresses a somewhat similar idea. He remarks, — "When we look at the system of isothermal curves surrounding the poles of revolution, and mark their coincidence with the magnetic poles of the earth, and their similarity to the isodynamical magnetic curves, we are disposed to view so remarkable a phenomenon, as the result of a physical condition of the earth itself, and produced by causes connected with its magnetic, or galvanic, or chemical agencies."

necessarily the producing cause of the convulsion which up-lifted them. Although we know that crystallization takes place at times without sensible heat, by the magnetic power, we are ready to admit that in the hands of God heat may have been the instrument of re-creation; but, because force under this form may have been His means of action, it does not follow that it now remains pent up in the bowels of the earth. The same force under other forms of power may be doing His will in the preservation of the globe which he fitted for the home of man.

But how little of the interior of the earth is known! Man has penetrated a few feet only. The upraised mountain which is supposed to indicate its deeper structure, may have been changed by its elevation, and may not represent the greater depth. We know nothing of the beds of the ocean, nothing of the internal condition of the earth. Scientific men in Europe are now disputing whether this globe is a hollow sphere, or a solid mass, or whether its interior is composed of molten fire. We venture to predict that the ultimate decision will not be in favor of the latter supposition. From analogy we believe in its perfect regularity of structure; the distortions and bendings of its upper rocks may be surface-changes only.

There can be nothing accidental in the position of any portion of the earth; and, when we recur to the fact, that for hundreds of miles the extended quarries of rock preserve one direction of dip, and one angle of stratification, — when we remember the order of the superposition of the various strata all over the world, we cannot regard the form of the earth as given to it by the accidental heaping together of heterogeneous masses of mineral, convulsively commingled from the bursting of its crust by the action of internal fire.

Science with its present knowledge has ascertained a more perfect order than chance explosions could produce.

In a previous chapter we alluded to the recently discovered magnetic properties of crystals. The experimentalist found that different crystals possess this virtue in varying degrees, and that, when suspended by delicate threads of silk, some point to the magnetic poles, and some to the poles of the earth's axis, while some act as declination-needles. The direction which the crystal takes is determined by the direction of its natural cleavages. Is not this earth, as it were, suspended and free to move under the impulse of magnetic force? Is not the very lay and dip of its mineral body, — the stratification of its framework of rock, so arranged that the flow of force may guide and determine its motion, as the minor stream of magnetic power influences the direction of the crystal? Its structure may have been given to it to adapt it for its required movements, and what appears to us as the accidental position of its parts, may be as much induced by law as the symmetry of the brightest gem.

CHAPTER XI.

"IN ALL OTHER SCIENCES THE PROPOSITIONS THAT WE ATTEMPT TO ESTABLISH EXPRESS FACTS REAL OR SUPPOSED; IN MATHEMATICS THE PROPOSITIONS THAT WE ATTEMPT TO DEMONSTRATE ONLY ASSERT A CONNECTION BETWEEN CERTAIN SUPPOSITIONS AND CERTAIN CONSEQUENCES." —*Stewart.*

It is supposed by many that the theory of universal gravitation has been mathematically demonstrated, and therefore stands incontrovertible. These questions will be asked: is not the mutual attraction of the heavenly bodies the very basis of mathematical astronomy, by which the position of every sphere is determined; and does not observation confirm the reasoning, and of course the theory on which the reasoning is founded?

The hand of a clock may point to the time of noon, and thus indicate the position of the sun, while neither the movement of the wheels, nor the movement of the pendulum which governs them, conforms in the least to the motions of the sun. Thus mathematical reasoning may track out the equable motions of the heavenly bodies, while neither the theory on which the process is founded, nor the process itself, in the least degree indicates the cause of the motion. A mathematical process is at times even aided by fiction, and is often successful when the theory on which it is based is known to be the very opposite of truth.

Under the Ptolemaic theory the apparent motion of the

heavenly bodies was considered the true and real movement. The earth was then believed to be the motionless centre around which the universe rolled. With this assumption the mathematician could, with surprising exactness, project eclipses, and predetermine the relative position of a sphere at any given future time. With this obviously false theory as the groundwork of calculation, results of great accuracy were obtained. This idea has been thus clearly expressed: "Now the truth is that this complicated and fantastic theory of the heavens, [the Ptolemaic,] with its operose contrivances of eccentric wheels, and circles riding on circles, and which in point of fact is false from beginning to end, is just as correct a basis for astronomical calculations, as the simpler, more beautiful, and more truthful system of Copernicus. The language of Mr. Whewell, whose authority on a point like this, no one will dispute, is, 'As a system of calculation it is not only good, but in many cases, no better has been discovered.'"

The character of the mathematical calculations, which are relied upon to prove that the theory of gravitation is applicable to the bodies composing the solar system, may be learned by the following quotation from the History of the Inductive Sciences: "The difficulty of doing what Newton did may be judged of from what has already been stated, that no one with his methods has yet been able to add any thing to his labors; few have undertaken to illustrate what he has written, and no great number have understood it throughout. The extreme complication of the forces, and of the conditions under which they act, makes the subject by far the most thorny walk of mathematics. It is necessary to resolve the action into many elements, such as can be separated; to invent artifices for dealing with each of these, and then, to recompound the laws thus obtained, into one common conception. The moon's motion, for instance, cannot be conceived without comprehending a scheme more complex than the Ptolemaic epicycles and eccentricities in their worst form; and the component parts of the system, are not mere geometrical ideas requiring only a distinct apprehension of relations of space to hold them securely; they are the foundation of mechanical notions, and require to be grasped so that we can apply to them sound mechanical reasonings. Newton's successors, in the next generation abandoned the hope of imitating him in this intense mental effort; they gave the subject over to the operation of algebraic reasoning, *in which symbols think for us* without our dwelling constantly on their meaning, and obtain for us the consequences which result from the relations of space and the laws of force, however complicated be the conditions under which they are combined."

Again; "if the mathematical calculations of the unmixed effect of the central force required transcendent talents, how much must the difficulty be increased when other influences prevented those first results from being accurately verified, while the deviations from accuracy were far more complex than the original action! If it had not been that these deviations, though surprisingly numerous and complicated, were very small in their quantity, it would have been impossible for the intellect of man to deal with the subject; as it was, the struggle with its difficulties is even now a matter of wonder."

"And even the possibility of doing what has been done, depends on what we call accidental circumstances, — the smallness of inclination and eccentricities of the system, and the like. If nature had not favored us in this way,

La Grange used to say, there would have been an end of geometry in this problem."

With a full admission of the accuracy of the results of mathematical reasoning as applied to astronomy, it does not follow that the spheres attract each other. An aberration or perturbation of the earth, calculated from the increased nearness of the sun, may be correctly calculated, and yet the aberration may not be occasioned by any attractive power in the sun. The motion of the earth may be changed, because of the greater force of motion relatively to the diameter of her orbit. The navigator finds the position of his ship by the aid of the imaginary lines of latitude and longitude intersecting each other; but it is not these lines which give position to the ship. Thus may the astronomer find the place and determine the motion of a planet by imaginary forces, — the one acting inward, the other outward; but these forces do not give position to the star. His calculation might be more difficult, yet no less certain, if in the place of two conflicting forces he used as the basis of his reasoning but one force giving directly the form of the orbit.

The arithmetical calculator takes for truth the simple relations of numbers. He believes in his data intuitively; and he extends these, the more simple relations, to the more complex ratios and values. The astronomical mathematician assumes that bodies fall by the attraction of the earth, and he extends the attraction of matter to the spheres. This extension partakes not only of the uncertainty of the premises, but also of the uncertainty which may arise from errors of mathematical reasoning. The assumption may be gratuitous, and the process, thus intricate and complicated, may be defective. Uncertainty from two sources, therefore, attaches to his conclusions. The results of this mathematical reasoning come not with the authority of the multiplication table. There is a vast difference between pure mathematics and the abstrusities of reasoning which extend mere speculation, by symbols that think for the calculator. The assertion that two and two are four, makes a different impression on the mind from that made by the calculations, by which the dynamical laws of nature are mathematically handled. Number, quantity, ratio, constitute the domain of mathematics. In these it is absolute, while it is the aid, the coadjutor only, in natural philosophy.

We quote the following from Stewart's Philosophy: "In pure mathematics when the truths we investigate are all coexistent in point of time, it is universally allowed that one proposition is said to be in consequence of another, only with a reference to our established arrangements. Thus all the properties of the circle might be as rigorously deduced from one general property of the curve, as from considerations derived from the radii. But it does not follow that all these arrangements would be equally convenient; on the contrary it is evidently useful and indeed necessary to lead the mind as far as is practicable from what is simple to what is more complex. The misfortune is that it seems impossible to carry this rule universally into execution, and accordingly in the most elegant geometrical treatises which have yet appeared, instances occur in which consequences are deduced from principles more complicated than themselves."

Every one in the least acquainted with mathematical astronomy is aware that its great burthen has been the process of correction. Thus the mass of Jupiter, his attractive value, has been calculated and recalculated; and is it wonderful that in all the changing results a value should at last be hit upon that would bring his influence under the theory of gravitation? The operation by which such results are obtained, involves the process of integration, "which cannot be performed in an immediate manner, since the quantities to be operated upon depend upon the facts, and thus require us to know the very thing which we have to determine by the operation. The result must be got at therefore by successive proximations. We must find a quantity near the truth, and then by the help of this, another, and so on." Whewell says, "The form in which the question of the truth of the doctrine of universal gravitation now offers itself to the mind of astronomers is this: that it is taken for granted that it will account for the motion of the heavenly bodies, and the question is with what supposed masses it will give the best account." This is the same as if it were asserted that twenty plats of ground contained an equal number of acres, and, this being taken for granted, the surveyor's question is, what assumed angles of each plat will best conform the theory to his calculations of the areas?

But if the mathematical process of thought could embrace the laws of force and motion,—could extend the law of falling bodies to the mutual relations of the planets, there is yet no determinate law under the theory of gravitation to be carried up from the earth. The law of the uniformly accelerated motion of a falling stone has no application whatever to the equable motion of the heavenly bodies, and the law of uniformly accelerated motion is the only principle which the fall of bodies to the earth has yet established.

The law that attraction is in direct proportion to the mass of matter which attracts, and in inverse proportion to the square of its distance from that which it attracts, can never be proved from the action of a falling body. Bodies falling to the earth, without reference to the medium in which they move, fall in the same time; a heavy pendulum vibrates in the same time as a light pendulum with the same arc; the world is so large in relation to falling masses, that the reciprocal attraction cannot be observed, and all that is known of the earth's attraction is, that all falling masses, whatever be their quantity of matter or their distance, are drawn with equal force. By the theory, indeed, there is an ounce weight's difference between the strength of the earth's attraction for one ton at her surface, and her attractive force for this ton a mile above her surface; but this is too small a diminution of attraction to be the subject of accurate experiment. To test, therefore, the theory of attraction relatively to mass and distance, we must go up from the earth to the spheres. There again is another defeat; for, though the relative distances of the heavenly bodies can be calculated, there is not one of them of which the mass, density, or attractive power can ever be known; and without this knowledge there is no proof that they do attract according to mass and distance. They attract as much as — they do attract!

"It is by the means of the perturbations of the planets," says an eminent astronomer, "as ascertained by observation, and compared with theory, that we arrive at a knowledge of the masses of those planets, which, having no satellites, offer no other hold on them for that purpose." . . "The perturbations of the satellites of Jupiter have led, and those of Saturn will doubtless hereafter, lead to the proportion of their masses to their primaries." In this view, all difficulties are avoided; the spheres do attract according to the law of mass and distance, for the mass is of a changeable value to suit the theory of attraction. Thus, the earth's diameter is, say eight thousand miles, the moon, say two thousand miles; the mass of the earth being one, the moon to fit her for her attracting office, is placed at 0.125172. Bulk and distance are of no consequence whatever; for the mass of any sphere is so fixed as to make all things conform to a law, which is not indicated by experiment or accurate observation. But even now there are difficulties, and ever have been. Says Herschell, "the curious and complicated effect of perturbation has given more trouble to geometers than any other part of the lunar theory. Newton himself had succeeded in tracing that part of the motion of the apogee which is due to the direct application of the radial force, but finding it only half what observation assigns, he appears to have abandoned the subject in despair. Nor when resumed by his successors, did the inquiry for a very long period assume a more promising aspect, and *strong doubts were excited whether this feature of the lunar motion could be explained at all on the law of gravitation.*" It was afterwards settled by taking *more properly* into account the tangential force.

There was, too, a difficulty with the secular acceleration of the moon's mean motion, "which, like the great equation of Jupiter and Saturn, had been long the subject of toilsome investigation. Some were again on the point of declaring the theory of gravitation inadequate to its explanation, and others were for rejecting altogether the evidence on which it rested." In this dilemma, La Place stepped in to rescue astronomy from its reproach. There was a difficulty, too, in some of the relative motions of Jupiter and Saturn, "*at one time too hastily regarded as almost subversive of the doctrine of gravitation.*"

Not many years ago doubts were entertained by the German astronomers, whether the law of gravitation were rigorously true with regard to the planetary bodies. Some calculations showed that the attraction of Jupiter as manifested by the perturbations of the small planets, Juno, Vesta, and Pallas, was different from the attraction on his own satellites. They inclined to the opinion that the attraction of the spheres might be *elective*. But this was met by a new determination of the mass of Jupiter by Mr. Airy.

One thing is certain, that it was a work of great time and labor, so to arrange the masses of the spheres, and so to construct the mathematical theory, that the perturbations could be made to conform to the hypothesis of gravitation; and that now the mathematical theory is far in advance of observation, bearing in a degree the same relation to perturbation, that the mathematical theory of the tides bears to the observed facts of the tides. There is perhaps much to cause the astronomer who believes in gravitation, to believe also that perturbation is caused by mutual attraction; and yet these perturbations could not of themselves, apart from theory, have led to the theory, nor can they prove the theory of themselves, for the reason that the masses of the spheres are determined by the theory itself.

We form the most definite idea of perturbation by the analogy before alluded to, presented by a distinguished astronomer: — "The laws of perturbation will be indicated by supposing all the spheres to be connected by imperfectly elastic material ties. By their motion there would be a forced vibration or oscillation communicated through the whole system of worlds." Perturbation is periodic and diffusible; the amount of eccentricity is equally divided, — what has been called the "eccentricity fund" is the property of all in a certain fixed ratio. There is nothing erratic, incidental, accidental about it. The periods of the sidereal year of the earth will be the same after a thousand more revolutions. The mean form of the orbit will never change. It is independent of the number of worlds, and of the quantity of matter that is or has been on either side of it. Whatever be the tie of the worlds, astronomers should never have applied to fixed and periodic changes the word "perturbations." The word does not belong to the movement of worlds, so exactly balanced by Infinite Wisdom.

In what other respect than as accounting for perturbation is the theory of gravitation even remotely connected with the movements of the heavenly bodies? Not surely with their motion. That gravitation gives motion to the planets has never been assumed by those who invented the theory, nor by any of the distinguished philosophers who have upheld it. All acknowledge an original impulse of motion. It is supposed that matter, when once set in motion, tends to move onwards *in a straight line* with a uniform velocity forever. This assumption that all motion is naturally in a straight line, is, we believe, the foundation error; and it is this assumption which necessitates the supposition of an attractive force to give to the straight line its *unnatural* curve. The assumption of this necessitates the supposition of the centrifugal force, to keep the planets from falling inward when drawn by the attractive force. Now it is found that the curvilinear motion of the planets can be mathematically represented by the action of these two supposed forces, — the one, of gravitation, which constantly deflects the straight line into a curve, — the other, of repulsion, which prevents the planet from being *too much* deflected. From these, though they may be mere fictions, can be deduced mathematically the true motions. Of course, this success of the mathematician does not in the least prove the theory.

Whewell says, "many men of good intellectual culture entertain in relation to gravitation vague and perplexed ideas, which show very clearly that the acceptance of the idea, is the result of traditional prejudice, not of rational conviction." If the theory of gravitation be true, it is indeed time that every educated man should have fixed and definite ideas concerning it. It is not claimed for the theory that this earth falls toward the sun as a stone falls to the earth. Attraction is equipoised by the repelling force. It is a statical force only. With the aid of its opposite force, it preserves the earth's distance from the sun. It is thus supposed only to determine the curve of the orbit; and this curve is modified somewhat by the attraction of other bodies. This is the extent of all that can be claimed for gravitation as acting in the solar system.

We have then this simple intelligible statement: the impulse of the motion of the planets which, not modified, would produce motion in a straight line, by the action of the centripetal and centrifugal *tendencies* results in the curvilinear. The attractive force is entirely incompetent of itself to give the orbit. It requires the supposition of an opposite force of equal intensity. If we should begin with a belief in the *repulsive* power of matter, this belief would require also the belief of an opposite force of equal intensity. Each fiction requires another fiction,—the one destroying the power of the other. They are useless. The planet with them, or without them, moves by its own impulse of motion.

If the centrifugal force is the tendency of a planet to assume a straight line of motion, the strength of this force must be determined by the strength of the impulse of motion. If the planet's force of motion is always of one degree of intensity, so must be the centrifugal force. But a varying attractive force requires a varying centrifugal force. The curve of the orbit demands an equal balance of repulsion and attraction. If the sun attract the more, centrifugal force must repel the more, and if by a greater distance the sun attract the less, centrifugal force must repel the less, to preserve the curve.

The velocity of a planet is determined by its distance from the sun. Unbalanced attraction and repulsion cannot give a determinate distance, — balanced attraction and repulsion neutralize each other at any distance. We believe, then, that we must go at once for the curve of the orbit, not to an original impulse of motion, but to the intensity of the force present to the planet. It is this, and this only, that determines the distance, the curve of the orbit. It is the intensity of the impelling force which governs the revolution of the spheres. Mathematicians, for the purposes of calculation, may refer to attractive and repulsive forces, may resort to any fictions or hypotheses which aid their processes; but in philosophy these imaginary lines can at once be discharged. We may look to the heavens with more simple and comprehensible ideas. We may clear the starry hosts from all mathematical intricacies, and no longer associate the planets with radial and tangential forces, and the complex fictions of human weakness; we may regard the majesty of their movements as true to the one great law, — velocity in proportion to orbit, distance according to the impelling force, — the spheres as united and oscillating as a great whole by the sweep of the bands of power, yet each pursuing its own course by its own inherent energy, centuries upon centuries passing over them, yet in all their complex relations no permanent change.

We do not mean by these remarks to undervalue mathematical research. Mathematics has its power from its peculiar exactness of reasoning, and the definite language which it employs in its processes. Mathematical training gives the power of concentrated thought; but, as with all other kinds of reasoning, its results, whether beneficial to the world or not, depend on the character of the mind which employs it. It is not an end, but an instrument; its value consists not in the ability with which the instrument is used, but in the work accomplished by it. Connected with the physical sciences, it may uphold error as well as advance truth.*

But there is an error which has done much evil in the world; that is, that it requires the power of abstract mathematical reasoning to understand mechanical philosophy. It is not so. The truth, if there are truths only to be arrived at by the process of mathematical reasoning, will be furnished by the mathematician. His results, if valuable, can be translated into the popular language. Mathematical investigation is a process, not a result,—an aid, but not success. Ferguson, who was surely distinguished by his

* Stewart, in his Philosophy, deprecates "that unqualified application of the mathematical method to physics, which has been fashionable for many years, and which seems to have originated chiefly in the commanding influence which the genius of Leibnitz has so long maintained over the scientific taste of most European nations." He speaks "of the obvious tendency which it has to withdraw the attention from that unity of design which it is the noblest employment of philosophy to illustrate." . . . "The consequence has been, (in too many physical systems,) to level the study of nature in point of moral interest with the investigations of the algebraist,—an effect which has taken place most remarkably, when, from the sublimity of the subject, it was least to be expected, in the application of the mechanical philosophy to the phenomena of the heavens."

astronomical knowledge, hardly knew the notation of algebra, and could not demonstrate a proposition of Euclid; and Kepler, whose name is indelibly impressed on the planetary system, was not even a good arithmetical calculator. Philosophy is not thus fenced out from men of ordinary education and common sense. The truths to be understood are few and simple; and it would be of great service if the general principles of philosophy were examined by those who have no educational bias to any of the present theories; the whole, after all, is but an appeal to the common sense of those who are sufficiently educated to understand the subject.

It is admitted that gravitation cannot account for the form and orbits of comets; and, if not for these, then it cannot for the minor eccentricities of the planets. The difference is in degree only, not in character. Sir John Herschell, in speaking of a comet, says that the laws of gravitation are insufficient to account for such a form of equilibrium, and that such a form is inconceivable without the admission of repulsive, as well as of attractive forces. "But if we admit the matter of the tail to be repelled from the sun, and attracted by the nucleus, it no longer presents any difficulty." He hazards the opinion, in order to obtain a repulsive power, that "the sun is permanently charged with electricity." Arago found that the attractive force, as exercised on heavy bodies, is inadequate to explain some peculiar motions of comets; and is of opinion that these motions "indicate a polar force which turns one semidiameter towards the sun, and strives to turn the opposite side away from that luminary." To aid the theory of gravitation in its application to comets, the supposition of a retarding medium has also been advanced, but with little success.

The difficulty, as we have said, is not in the character of the motions of these bodies; but the great degree of the eccentricity of their orbits, shows plainly a defect in the theory of gravitation, which escapes notice when eccentricity bears but a small proportion to the diameter of the body or to its orbit of revolution. It is impossible not to perceive that the sun has not an attractive power, when we trace the motion of a comet. We will suppose a comet, millions of millions of miles away from the sun, moving from him under the impulse of its impelling force, until the central attraction, when immeasurably *weakened* by distance, changes the direction of the wanderer and recalls it toward the centre; it returns rushing toward the sun, almost in a straight line, and when so near that another quarter of an hour would bring it into contact with the central luminary, at this near point, although attracted to the centre with a force almost as great as that with which the sun binds together the parts of his own mass, it again speeds away against the full force of the sun's attraction.

Nor can the sun give form to the orbits of the planets; for he cannot, with a force invariable on the same mass and at the same distance, determine the form of the ellipses, or the degree of the eccentricity. Attraction does not shape the orbit. We are somewhat acquainted with the mathematical reasoning by which the unvarying central force is supposed to give the modifications. But, after all, an equal central force acting on any heavenly body at any one distance, cannot do aught but maintain that distance against the uniform centrifugal force, which is measured by the intensity of the impulse of motion. If there are changes of velocity and changes of curve, they must come from a change in the force which determines the motion. There is something belonging to the planet, not to the sun, — a change at the circumference, not at the centre of revolution, which gives the peculiarities of the orbit. The difference between the orbits of Mercury and Venus, for instance, comes not from solar influence; this influence varies only with the mass and the distance. It is not mass or distance that determines eccentricity. Therefore, that which gives a different character to the orbits of these two planets is local, and peculiar to each. The eccentricity of the orbit does not come from central attraction, nor does its elliptical form; nor consequently, does the mean distance, nor the curve, nor indeed any peculiarity of the orbit of any planet.

To our mind it is another disproof of the theory of gravitation as applied to the heavenly bodies, that it throws over the solar system a complicated irregularity. It takes away symmetry, the simple and orderly relation of the spheres; by the arrangement of masses to suit the theory it destroys the idea of regular progression. For instance, the theory presents the moon as a mass of vapor compared with its primary, while one of Jupiter's satellites is made more dense than Jupiter himself. One planet is puffed up, another unnaturally hardened, and relative density skips back and forth without method or order. Uranus is less dense than Saturn; Venus and Mars less dense than the Earth, which lies between them. As it has been most singularly expressed, some are as light as vapor, and some heavy as lead; and between the extremes are those of the consistency "of water, of honey, of antimony, and of pine wood," all confusedly intermingled for the purpose of adapting the heavens to the theory of gravitation.

To show that we do not exaggerate in these statements, we will quote from a standard authority. "The planetary system in its relations of diameter and relative position, in density, time of revolution, and eccentricity of orbit, does not offer to our apprehension, any stronger evidence of natural necessity than the proportion observed in the distribution of land and water on the earth, and the height of mountain chains. In these respects we can discover no common law in the regions of space, or in the irregularities of the earth's crust." And the heavenly bodies are presented "as facts arising from the conflict of forces acting under unknown conditions." We cannot avoid the belief, that no theory can be sound which does not unfold the beautiful symmetry and perfection of the universe.

It has been found that the diameter of the earth multiplied by one hundred and ten, gives the sun's diameter, and that this product, the sun's diameter, multiplied by one hundred and ten, gives the distance of the sun from the earth; and also that the moon's diameter multiplied by one hundred and ten, gives her distance from the earth:

$7920 \times 110 = 871,200$, observed diameter 880,000
$871,200 \times 110 = 95,832,000$, " distance 95,000,000
$2160 \times 110 = 237,600$, " " 237,400

It appears from this, that the diameters of the heavenly bodies have not only a fixed proportion to each other, but also to their distances one from the other. Diameters therefore are not, if we may be allowed the expression, deceptive; they have an honest value. It is the circumference of the planet and the sweep of its orbit, that determines its relation to others. Magnitudes as presented to the astronomer are more reliable than speculations as to density. The one, accurate instruments and close observation will give with great exactness; the other, however acute may be the mathematician, however sound his system of calculation, may be altogether erroneous.

We do not believe that it was by the "conflict of manifold forces," that a one-hundred-and-tenth part of the sun was torn off, and carried one hundred and ten times his diameter in distance, there to be consolidated and shaped under the direction of the centripetal and centrifugal forces; and that the accompanying vapors were rolled off so as to form a moon exactly one hundred and ten times its diameter in distance from the earth. We believe that the diameters of the heavenly bodies are measured one by the other, and bear to one other a fixed proportion; that the same measure was used, and was guided by the same hand all over the system of worlds; that all apparent disorder or irregularity is the distortion of facts as seen through the mists of our ignorance.

We will assume for Mercury a diameter of three thousand six hundred miles, and compare this with the diameters of Venus, Earth, the Moon, and Mars. If we discover a fixed relation in their magnitudes, we may reasonably suppose that the chain of progression extends through the whole system, even if some of its connecting links are hidden.

Mercury, ...		3,600	observed diam.		3,240
Venus, ...	3,600 + 3,600	= 7,200	"	"	7,700
Earth and Moon,	7,200 + 7,200	= 14,400	assumed value,		14,400
Earth, ...	7,200 + 720	= 7,920	observed diam.		7,920
Moon, ...	720 + 720 + 720	= 2,160	"	"	2,160
Mars, ...	2,160 + 2,160	= 4,320	"	"	4,180
		39,600 miles			39,600

The aggregate of our assumed diameters, as thus "double one against the other," is the aggregate of the observed diameters; and, while the aggregates thus agree, the diameters assumed for each do not vary materially from the magnitudes given by the astronomer. If we can show that the value of the moon by increased rotation is 6480, or in other words, that the joint value of the earth and moon is 14,400 miles, we establish the fact that the diameters of the planets are "built one upon the other."*

The velocity of the earth's rotation at the equator may be stated as 1,037 miles an hour. If the diameter of the earth were increased by the diameter of the moon, (7,920 + 2,160 = 10,080,) rotation in the same time would give a velocity of about 1,260 miles an hour. The velocity of the moon's rotation is 2,300 miles an hour. This is made up of the increase of 230 miles, which would ensue were she a component part of the earth, and of twice the earth's velocity, constituting the 2,070 miles. Thus, her rotation at the circumference of the local system is three times the value of the rotation of the earth at the equatorial regions.

Earth,	. . . .	7,920
Moon,	2160 × 3 =	6,480
		14,400 miles.

By the multiplication of 14,400 by 16, we have 230,400 miles, the minor axis of the moon's orbit; by the multiplication of 14,400 by 17, we have 244,800, the major axis. The eccentricity of the ellipse of the orbit is stated at $\frac{1}{17}$. The mean of the two extremes 237,600 is exactly the mean distance of the moon from the earth as deduced by observation, and is the distance as obtained by the multiplication of the moon's diameter by 110.

The distance of Jupiter is $5\frac{1}{2}$ times the distance of the earth from the sun. $14,400 \times 5\frac{1}{2} = 79,200$, which is the small diameter of Jupiter. If we multiply the moon's diameter by $5\frac{1}{2}$, we have 11,880, which is about the aggre-

* These "numerical coincidences" extend through the whole system.

gate of the observed diameters of Jupiter's satellites, namely, — 2508 + 2668 + 3377 + 2890 = 11,443 miles.

The mean distance of Mercury from the Sun is thirty-eight millions of miles, or forty-five times the Sun's diameter. Deduct one diameter, and 44 × 2 = 88 gives the revolution of Mercury in days. The moon is 16½ times 14,400 miles from the earth; 15½ × 2 = 31 days rotation. Is this a mere numerical coincidence? *

The *sidereal* revolution of the moon is, say twenty-seven days; the sun's rotation, as increased in relation to the earth by the earth's motion in the same direction, is twenty-seven days. In movements thus involved and combined, the sun and moon rotate the same number of times for every revolution of the earth. Are the form of the orbit, the circumference of the planet, and its motion, determined by its mass and distance according to the theory of gravitation? We rest in the belief of a perfect symmetry of all the parts of the solar system, and a close sympathy of motion constituting the universe a perfect whole.

The preceding are rough calculations, selected from many of the same character, not as the most striking, but because in a simple, unartificial manner, they present the fact that the diameters of the spheres, as it were, grow out of each other, and that there is a determinate relation of magnitude and distance running through the system of worlds. The

*Mercury, ...	45 × 2 =	90 days,	observ.	88 days
Venus, ... 45 + 22½ =	67½ × 3 =	202½ "	"	220 "
Earth, ... 67½ + 22½ =	90 × 4 =	360 "	"	365 "
Mars, ... 90 + 45 =	135 × 5 =	675 "	"	685 "
Vesta, ... 135 + 90 =	225 × 6 =	1350 "	"	1350 "

We thus roughly indicate a regular progression of the times of revolution, which can be extended through the system with striking numerical coincidences.

relations of these bodies to one other seem to demand the greater attention; for, when distinctly traced out, all seeming incongruities will vanish, and order will be restored to the heavens, or, in other words, the mind in sympathy with nature will discover the order that ever has reigned and ever will reign supreme.

From the sudden decrease of the diameter of Mars relatively to that of the Earth, analogy would lead us to believe in a satellite or satellites accompanying Mars. But the telescope does not reveal the existence of these attendants. What are the relations of the Asteroids to Mars?

By a hasty generalization we are disposed to end the first series of planets with the Earth, and to consider Mars as intermediate, or pertaining in some degree at least to the system of worlds between the Earth and Jupiter. Is there not another system of Asteroids between Saturn and Uranus, and another beyond the bounds of Neptune, and does not system thus interlace with system through the great whole?

We turn our attention for a moment to the nebular hypothesis, a short time since so generally believed, now fading away under the clearer light of modern science. It has been thus described: — "The assemblage of stars that form at present our solar system, were first confounded in one celestial body, resembling one of those mysterious nebulæ we see floating in the celestial spaces." . . . . "But soon the development begins. A principle of concentration, — gravitation, — counterbalances the unlimited expansion of the gaseous matter, brings the molecules nearer together, and groups them in a spheroidal mass. This approximation allows the molecules different in nature to act upon each other according to their chemical affinities; the process of life commences, and its earliest manifestation is light and heat. The nebula is detached from the general mass, under the form of a luminous spheroid traced in the obscurity of the heavens. This is the first step in the process of formation." "This gaseous spheroid then resolves itself into local agglomerations, which, while concentrating each in itself, under the influence of gravitation and chemical combinations, separate from each other in distinct spheres. Whether this phenomenon is effected, as La Place imagines, by the successive separation and agglomeration of concentric layers of the solar atmosphere, or in virtue of some organic law still unknown, is of little importance here. The fact of the separation of the different bodies of our solar system into a number of spheres, planets, and satellites, is not less certain, and constitutes one of the essential and incontestible phases of its development."

From another work we extract: — "The idea, then, which I form of the progress of organic life upon the globe, (and the hypothesis is applicable to all similar theatres of vital being,) is, that the simplest and most primitive type (under the law to which that of like production is subordinate) gave birth to the type next above it, and this again produced the next higher, and so on to the very highest." . . . "It has been seen that, in the *reproduction* of the higher animals, the new passes through stages in which it is successively *fish-like* and *reptile-like*." We make these quotations to give what is called the progressive theory as applied, not only to world-making, but also to the development of animal life.

To all these theories we may apply the language of Cudworth; we quote from his Intellectual System of the Universe, first published in 1678. "Wherefore, infinite atoms of different sizes and figures, devoid of all life and sense, moving fortuitously from eternity in infinite space; and making successively several encounters, and consequently various implexions and entanglements with one another, produced first a confused chaos of these omnifarious particles, jumbling together with infinite variety of motions; which afterward, *by the tugging of these different and contrary forces*, whereby they all hindered and abated each other, came, as it were, by joint conspiracy, to be conglomerated into a vortex or vortices; where, after many convolutions and evolutions, molitions and essays (in which all manner of tricks were tried, and all forms imaginable experimented,) they chanced, in length of time, here to settle into this form and system of things, which now is, of earth, water, air, and fire; sun, moon, and stars; plants, animals, and men; so that senseless atoms, fortuitously moved, and material chaos, were the first original of all things."

There is another theory somewhat more recently revived. It is the opposite of the development hypothesis, whether applied to world-making, or to the production of the higher orders and types of animal life from the lower. It is the very reverse,—a degradation theory; and shows us animal life sinking below its former level. The reptiles, forsooth, are not so large as they once were. There are species of fishes brought into existence with eyes misplaced, and mouths askew. We will again quote from Cudworth: "Wherefore, they affirm that the earth brought forth divers monsters, and irregular shapes of animals;

> "Orba, pedum partim, manuum viduata vicissim,
> Multa sine ore etiam, sine vultu, cæca reperta."

Of course, we do not impute these speculations to the theory of gravitation. They are not new. In the old atomic philosophy, long before the theory of gravitation was announced, they were presented. But it is an objection to this theory, that it permits the belief of the world-making and life-giving power of matter, as resuscitated in modern times, to find a resting place in the mind. If it does not produce the nebular, the progressive, and the degradation hypotheses, it does not place itself as a barrier against their introduction. This theory and these speculations are not in opposition to each other; and it behooves those who would maintain the doctrine of the attractive power of matter, so to limit this power that it shall never be represented as the potential cause of all things.

CHAPTER XII.

"IF HIS SUN ROLLS OVER MY HEAD AND WARMS ME, — IF HIS WIND COOLS AND REFRESHES ME, — IF HIS VOICE SPEAKS TO ME, WHETHER IN THE THUNDER AT MIDNIGHT, OR IN THE WHISPERS OF THE FOREST, OR BUT IN THE RUSTLING OF A LEAF, — IF HIS SEASONS STILL COME ROUND TO ME IN THEIR GRATEFUL VICISSITUDE, — AND WHEREVER I LOOK IN OUTWARD NATURE, I BEHOLD CONSTANT ACTION, CHANGE, AND JOY, — I DO NOT SUPPOSE THAT BRUTE AND SENSELESS MATTER CAUSES ALL THIS BY ITS INHERENT POWER, WHETHER ORIGINAL OR DERIVED, — BUT THAT THE SPIRIT, THE PERSON WITHIN, CONTROLS, VIVIFIES, AND PRODUCES ALL."

Francis Bowen.

IT may be said that if our assertion, that the earth is subject to occasional and periodic oscillations of surface, were correct, these oscillations would be indicated by cracks and fissures; that as earthquakes show their agitations, so would these fluctuations declare themselves by inequalities of surface. But this would not follow; for earthquakes are limited, abrupt disturbances, while the more gentle undulations of the earth, which we suppose, would be analogous in their movements to the bending of an elastic spring, the curve of which is too gradual and too extended to produce any rupture in its parts. Elevation of the earth's surface would gradually fade away into depression without any break or abrupt change to mark its bounds.

The Britannia tubular iron bridge, it is said, at times contracts and expands to the extent of three inches. This is a much greater change of atomic structure in proportion to the length of the bridge, than the change in the structure of the earth necessary to produce the rise and fall indicated by the barometer. "The action of the sun at mid-day does not move the tubes of the bridge more than a quarter or three eighths of an inch. The daily expansion and contraction of the tubes varies from half an inch to three inches, attaining either the maximum or minimum about 3 o'clock, A. M. and P. M." This statement is copied from a London paper; if correct, it is in some degree a confirmation of our views.

Unquestionably there are many plausible objections that can be urged against the hypothesis which we have offered. It is of course incomplete; there is much to be learned; much remains for examination, much for further explanation. We but indicate the mine, presenting only specimens of the surface-ore, not the raised and refined metal. The laws of force require assiduous attention for their development, as the currents of its transfer must necessarily be numerous and complicated.

Our system has not the simplicity of the theory that all matter attracts according to one unvarying law, nor is it so readily and easily applied. And for this very reason we believe that our explanation is more true to nature; for, while it throws a clear light over some prominent phenomena, it has not the vagueness which would permit its application at once to every fact. On the other hand, a theory may cover a vast ground from its very vagueness and indefiniteness. A loose and unshaped robe can be thrown over any form; but this easy adaptation does not prove that it is a fitting, useful garment. A theory is not proved to be true by the readiness of its application. The truth on the other hand, because of its exactness, shows its adaptation only where the fact to which it is applied is well understood, and its application can be extended just in proportion as the relations of things one to another are distinctly traced out. A true theory will be a theory of progress. Its reception will be marked by a constant advance in knowledge.

Suppose it were asserted that all matter was originally created in one dense and compact mass, and that then a repulsive power was given to it, increasing in intensity in direct proportion to distance and in inverse proportion to mass; or that space has no fixed and definite extension, but that it contracts and expands with an intensity in proportion to the mass and distance of the bodies which it separates. All facts could be at once arranged under these absurd hypotheses. They merely assert that things are placed as they are placed. Does the more reasonable theory of gravitation give any brighter light? Is it not characterized by the same vagueness, and the same readiness of application to all the phenomena of the universe?

In former times chemistry divided all matter into four elements, earth, air, fire, and water. This is a simple classification, intelligible and applicable at once. Had it been retained, however, the science could have made no possible advance. It therefore soon abandoned this easy generalization. In its progress it has divided and subdivided, and it now presents us some sixty elements, and these combined and separated by the conflict of manifold forces. But it will not remain a confused collection of facts. It will be reconstructed on a sure foundation. It will by its minute exactness advance to general principles, and while it develops the relations of atoms, and of the constituent elements of masses, it will discern the unity of force, and discover the general dynamic laws which have remained hidden because philosophy has supposed that already, in the law of attraction, it has attained its ultimate knowledge of the properties of matter.

There is some reason, therefore, to question a theory of too ready application, which does as much for the advance of science at first, at its very inception, as after the lapse of years. A theory may be apparently useful in collecting all minor doubts into one general doubt, — in removing obstacles from the by-paths of science, while it gathers together all those obstacles and forms a barrier in the highway of progress. The leading fact, on which the law of gravitation is founded, is the fall of bodies to the earth. To this fact it owes its origin. Of course, if it can explain aught with exactness, if it can give clear ideas in relation to any phenomena, — if it have strength, perspicuity, distinctness of illustration, unquestionable adaptation, here it will be found. And if in its application to a falling stone the mind receives no clear light, if even this fact remains enveloped in a mist of obscurity, surely there is some reason to question the value of the theory in its more remote and extended applications. We will now examine the theory of attraction as explanatory of the fall of bodies to the earth.

It is said that falling bodies move faster and faster the longer they continue in motion. To explain this accelerated velocity, we are told that "the force of gravitation acts not at once, but at every period of descent; of course the velocity increases as the time increases; hence the space described must be as the time multiplied by the velocity, which is the same thing as to say that the space described is as the square of the time, or the time multiplied by itself."

Now is it invariably true that falling bodies move faster

and faster the longer they continue in motion? Does the force of gravitation act, not at once, but at every period of the descent? The answer must be that this is not rigorously true in relation to bodies falling in an arc, as in the case of a pendulum, nor in the case of water in downward motion over an inclined plane. The space described by a falling body is as the square of the time, only when the fall is perpendicular.

It is the change of the level of rotation which gives the uniformly accelerated motion. The velocity has no relation to the time of the fall, nor to the space described by the motion. It is measured by the degree of the change in the orbit of rotation. This is the only statement rigorously true, and universally applicable to all the circumstances of a falling body. If we divide the space of the perpendicular fall into equal parts, the velocity of the first part will be as one, of the second as three, of the third as five, and so on; for the force of any part of the descent will be its own spare force of descent, with the addition of the force of the preceding parts. The same degree of force always describes the same space in the same time; therefore in the perpendicular fall the product is the same, whether the time or the space is squared.

The question then is, whether the uniformly accelerated motion of a falling body is the result of the time in which it has been moved by the gravitating power, or whether it is the result of the change of the level of rotation.

On our theory, the spare force of rotation being determined by the degree of the change of level, the downward motion of the falling body would have an increasing velocity; for the spare force increases with every line of actual descent. However minute the division of the space, the force at every division and the velocity at every division would be doubled. Hence the reason for squaring, or multiplying by itself, the space described by the fall. It is just what would be affirmed as the result of the theory, without recourse to observation, experiment, or admeasurement.

On the other hand, a uniform acceleration of the motion of a falling body could never have been supposed as a necessary result of the attraction of the earth, and when observation presents the fact, the law of gravitation can give no explanation or elucidation of it. According to the theory, the power of attraction varies only with the mass and with the distance. The accelerated motion is from an increased force without an increase of the attracting mass, and a force increased far beyond what the lessening distance could supply.

Most unquestionably, an increasing velocity of motion under the same circumstances implies a greater producing force. To obviate this difficulty, it is supposed that the attraction of the earth is cumulative, increments of force being added in proportion to the time in which the earth attracts. Why then does not the earth's attraction accumulate in the body when at rest, giving it more and more weight or *tendency to descend?* Why does not this attraction accumulate in water falling over an inclined plane? The water is in motion uniformly and constantly towards the earth. How does the earth, when putting forth an increasing attractive power, discriminate between bodies at rest, bodies moving obliquely downwards, and those in direct motion toward her centre? How does the earth proportion the velocity to be added to that already conferred? The theory of gravitation seems to require an intelligent attracting power, with a nice perception of time, and the ability of making mathematical calculations, so that force shall be added in the right degrees, according to the changing circumstances of the moving body.

Galileo, who first announced the law of the square of the time, was much perplexed by it, and ascribes it, somewhat doubtingly, to the simplicity of the works of nature, saying, "If we attend carefully, we shall find that no mode of increase of velocity is so simple as that which adds equal increments in equal times; as the uniformity of motion is defined by the equality of spaces described in equal times, so we may conceive of the uniformity of motion to exist where equal velocities are added in equal times." Had this train of reasoning been followed out, the acceleration would have been imputed to the change of level, to the distance of the fall, not to the length of the time of the fall; it would have been traced to the body falling, not to the earth on which it falls, — to a varying force, not to the uniform force of attraction. But this reasoning could not be resorted to by those who believe gravitation to be the cause of motion; therefore to this day remains the singular explanation of the uniform acceleration of falling bodies, that, to bodies in motion the earth adds constantly increasing increments of force to be measured by the squares of the times; that is, the power of attracting increases with the use of the power.

From this it follows that there are two differing attractions, the attraction which gives weight and that which causes motion, — the one of stationary intensity, the other increasing with the time of its action. A stone on the summit of a mountain has been attracted for ages without an increase of weight; when it falls how rapid is the increase of the power which draws it downward!

How simple and intelligible is the idea that the earth has no influence whatever in the fall of a stone, — that she imparts neither a uniform impulse nor constantly increasing impulses, — that the force is always, invariably, in the body which moves, — that it is the *extent* of the fall, not the time of the fall, which measures the extent of the force liberated from the force of rotation, — that this force according to its intensity gives motion, not downward only, but laterally and upward, — that, if the space described between the two levels of rotation of the descending body is measured, the quantity of force liberated to act in a new direction is ascertained. This simple solution takes away all the mystery which so long has enveloped the subject.

And the converse act, the raising of a stone to a superior level of rotation, confirms this view. To a stone at rest, additional force for its new relative position must be applied, and, however supplied, its intensity is just the force of its fall through the same distance; in fact, the force of the fall is the force for the rise, as is shown in the oscillations of a pendulum.

But what is gravitation? "A name," said one of our strongest men, "that is given to the veil which covers the unknown from our sight." Its original signification was heaviness. Transferred from the tendency to descend to the act of descent, from the act of descent to the effect of descent, from the effect of descent to the principle by which all bodies descend, thence it went from the earth to the heavens, first deflecting or bending the otherwise straight line of the motion of the moon, then curving the orbits of all the planets, then serving as the impelling power of the sphere, then as the cause of all the perturbations of the orbits of the planets, and of the eccentricity of the orbits of the comets, then causing the motion of the binary stars, one around the other, and wheeling the astral systems around a centre in the direction of the Pleiades. Returning again to the earth, attraction takes under its ample folds every act of masses, every motion of atoms, — the first, last, and only cord by which philosophy ties the creation together, in atoms, masses, worlds, and systems of worlds, aided perhaps somewhat by centrifugal force, — the abhorrence of nature for curvilinear motion!

The gradual extension of the power of gravitation is well known by those who pierce

> "the long-neglected, holy cave,
> The haunt obscure of old Philosophy."

Long before the days of Newton, the doctrine of the attraction of the earth was taught and generally believed. Magnetic attraction was the matrix of the idea. An analogy was early traced out between the descent of falling bodies and the rushing of particles of iron to the magnet. Thus was established a supposed principle, — a tendency or desire of bodies for the presence of each other, an idea which is thus curiously expressed by Bacon: — "The electrical operation, concerning which Gilbert and others since him have made up such a wonderful story, is nothing else than the appetite of a body excited by friction." This appetite of matter for matter was the germ of the theory, which has taken such rapid strides from its inception by the elder philosophers to the present conception of the schools.

The great extension of this appetite of the earth for the matter of adjacent masses to an appetite for the matter of the moon, dates from the time of Sir Isaac Newton. Before his time it was hinted at, supposed, timidly expressed, *considered possible.* The idea had never been distinctly avowed. It was in men's minds a "shadowy belief." Its destined enunciation was reserved for after-times. Newton, as we shall see, disclaimed the belief.

The first conception that the attraction of the earth extended to the other spheres thus took its rise. When a body is falling to the earth, it also partakes of the motion of the earth around its axis. Hence the idea that, some how or other, the fall to the earth is connected with the motion of the earth. The fall of an apple from the mast-head of a ship, which partakes also of the motion of the ship, and between the two forces describes a curve in its fall, is one of the connecting links of the fabrication of the chain of attraction, — both motions in the same time, one modifying the direction of the other, — surely they indicate one cause!

The idea had, however, its refining process, its transition state, before being fully incorporated into philosophy. "A ball thrown from a cannon describes a curve by the mutual action of the impelling force and the force of attraction. The explosion of the gunpowder would of itself give the rectilinear line, the curve is therefore the action of gravitation." The next step was, "If the impelling force of the explosion were sufficient, it would so far overcome the force of gravitation, that the ball would describe an orbit round the earth, without descending to its surface; if the impelling power were sufficiently great, gravitation would act only in giving the cannon ball an orbit of rotation round the earth like the orbit of the moon." Then came the mathematical demonstration, that an arc of the *deflection* of the moon's orbit, cut off for the calculation, is equal to the fall of a stone for an equivalent period of time. This settled all doubts, and the principle of gravitation, going up to the moon, from this elevation could easily soar higher to cover all the motions of the heavenly bodies. There is no hindrance now to any further extension, because, attraction being according to the density, as well as according to the distance of any sphere, philosophy can give to any of the heavenly bodies just the required power. Assume the law, then most easily can density be adjusted, so as to bring the power of the attraction of any star to the limits necessary for the theory. For instance, the sun has too great attractive power according to bulk for the theory, but what of this? His density can be arranged so as most exactly to suit the theory. For this reason no sphere can trouble the theory, and difficulties are met with only in comets; for in their case there is too much indistinctness, neither the bulk nor the mass of these bodies being known.

It is generally believed that Newton mathematically demonstrated that the attraction which impels a falling stone, gives the deflection of the orbit of the moon, — that is, admitting that the impulse which gave motion to the moon, was at the beginning an impulse of rectilinear motion, that the attraction of the earth gives the curvilinear form, and determines the circle of her revolution. We copy an abstract of the demonstration from the London Encyclopedia, Article, Newtonian Philosophy, as in a very few words it conveys the process. "If we imagine the moon deprived of all her motion, and be let go so as to descend toward the earth with the impulse of all that force by which it is retained in her orbit, it will in the space of one minute of time describe in its fall $15\frac{1}{4}$ Paris feet; for the versed sine of that arc, which the moon describes in the space of one minute by its mean motion, at the distance of sixty semi-diameters of the earth, is nearly $16\frac{1}{12}$ Paris feet. Wherefore, since that force in approaching the earth increases in the reciprocal duplicate proportion of the distance, and upon that account at the surface of the earth is 60×60 greater than at the moon, a body in our regions falling with that force ought in the space of one minute of time to describe $60 \times 60 \times 15\frac{1}{4}$ Paris feet, and in the space of one second of time to describe $15\frac{1}{4}$ of those feet, and with this very force we actually find that bodies here on earth do actually descend."

This demonstration is incomplete in many points of view. In the first place, *it is assumed* that the stone falls by attraction. In the next place, the moon being of one quarter of the diameter of the earth, there is her reciprocal attraction to enter as an element of calculation; for the attractive force of the stone and earth would not measure the joint attractive force of the earth and moon. Her mass, never to be ascertained, would influence the result. Again, the resisting medium of the motion of the stone and the moon may be different, making the fall and deflection of different values for the same time.

But admit that the stone falls by attraction, that the moon's greater mass does not vary the strength of the attractive force, and that the media through which they move are the same, the attraction for the stone is a constantly increasing force, giving constant and uniform acceleration of motion, while the attraction for the moon is a statical force, bearing with uniform intensity or weight. Attraction for the falling stone is a cumulative energy, — for the moon a constant power. The moon is not drawn toward the earth, but keeps the same orbit; she preserves, so far as relates to the force of gravitation, one distance. If two, three, or four similar arcs be cut off, of two, three, or four minutes, the aggregate of deflection will be found by the simple addition of the deflection of each, — the aggregate of the fall of the stone for two, three, or four seconds, will be found by squaring the time, not by the simple addition of the time.

If we take for one element of calculation an uniformly accelerated velocity to be handled mathematically in comparison with an equable velocity, it must be motion for a definite time, and the mean of this varying velocity must be deduced as its measure or value; for instance, take the time of the fall of the stone equal to one revolution of the moon, find the mean velocity, and compare this mean velocity with the velocity of the moon in her orbit. Will a body falling for the whole time of the revolution of the moon, have a mean velocity equal only to the deflection of the moon in one orbitual circle? Even here there will be only an approximation to a common standard; for the true element for calculation is the mean velocity of the falling stone for the whole duration of time for which the moon has been in motion. As the proposition has been demonstrated, the result is true of one given arc of deflection, compared with one second of the time and the resulting velocity of the falling stone. The moon, if she were let go and fell to the earth, would unquestionably obey the laws of falling bodies. Her spare force of rotation would be in the same ratio according to the change of level. Every line of her downward motion would increase her velocity. Continuing in her orbit, her motion is uniform, equable, ever the same, according to her distance from her centre of rotation around the earth or its axis. The two motions have no analogy; there is between them no ground of comparison; the one is equable, the other accelerated, — the one is fixed, the other without bounds, — the one is incidental, the other perpetual. They cannot be bound together in the same mathematical reasoning, save under the law of equal areas, and that by which decreasing areas of motion give out the same force that equal increasing areas of motion would require.

To combine them demands that a retarding element be added to the one, or an accelerating element to the other, and the degree or value of this new element cannot be mathematically expressed, since it involves the conception of an infinite series. It plunges us at once into the depths of transcendental mathematics.

Newton was aware of the incompleteness of the demonstration. He believed in a subtle ether which filled all space, and which gave to the spheres a retarded velocity. His disciples discarded the ether, and succeeding philosophers adopted the bold and bare conception of gravitation, which Newton thus expressly disavows in his often quoted letter to Dr. Bentley.

"It is inconceivable," writes Newton, "that inanimate brute matter should, without the mediation of something else which is not material, operate upon and affect other matter without mutual contact, which it must do if gravitation in the sense of Epicurus be essential and inherent in it, and this is the reason why I desire that you would not ascribe it to me. It is so great an absurdity that I believe no man, who has in philosophical matters a competent way of thinking, can ever fall into it." * So even those may dis-

* On this subject, Stewart remarks in his Philosophy, — "It is impossible to conceive, in what manner one body acts on another at a distance, through a vacuum. But I cannot admit that it removes the difficulty to suppose, that the two bodies are in actual contact. That one body may be the efficient cause of the motion of another body placed at a distance from it, I do by no means assert; but only that we have as good reason to believe that this may be possible, as to believe that any one natural event is the efficient cause of another."

card the attraction of brute matter most who revere the name of Newton, — those who are

> "passionate for ancient truth
> And honoring with religious love, the great
> Of elder times,"

are not compelled to adhere to a theory, which gives one sphere a power over another sphere, at a distance from it, without the mediation of something which is not matter that passes between them.

It is not wonderful that Newton repelled almost indignantly the imputation that he believed in the attraction of brute matter; but it is surprising that any man should have a distinct conception of the harmony of the spheres, of their equable unchanging velocities, of the fixed law of their movements, and find in their uniformity an analogy with a falling stone, between which motions there can be no relation whatever, no ratio of velocity, no common condition, — the one being permanent, the other incidental; the one, as it were the accident, the other the law of nature; the one, a flowing in a determined orbit, the other a passage from one orbit of rotation to another orbit of rotation; the one bearing the stamp of perpetuity, the other being the evidence of change; the one continuous, the other enduring but for a time. Equally surprising is it that any one can look upward to the revolving spheres, in number exceeding the sands of the sea-shore, all, every one, moving in curves; or look downwards, seeing there no straight lines of motion, — the smallest level of the earth to the eye of philosophy being a curve, — knowing that the straight line, wherever drawn, would be a line of collision and impingement, — and yet affirm, that not one sphere for one moment could be held to the curve, without the conflicting force of gravitation to overcome the natural rectilinear law of motion.

We have thus passed hastily over a vast extent of illustration, and present this rough outline to the public eye. We wish that it were more worthy of the subject, more faithful to the idea that we have formed.

Not by way of apology, but to give the reason for it, we will advert for a moment to our frequent recurrence to the religious bearings of science, and to the Mosaic account of the Creation. The association of this research with the Bible, came early with the thought that in the instance of the rainbow philosophy was but the echo of the voice of God, and the connected passages were read with an increased reverence and intense interest.

There is no man, however skeptical the usual frame of his mind, however careless and worldly the usual train of his thought, who does not at times look upward for support and direction. Nor, in a scientific inquiry, when baffled in the attempt to comprehend the works of Nature, when tossed on the changing wave of theory, can the student refrain from seeking light, direction, authority, from above. The mind is so constituted as to look for some revelation beyond what the face of nature presents.

Were the Mosaic account of the beginning but the freshness of philosophy, the earnest expression of the human mind, when the earth in her early morning was first examined by man, — were the ideas mere thoughts handed down from the beginning of the world, — mere traditions kept pure and uncontaminated in the Bible, preserved from the defilement of a passage through the minds of countless generations, — like the pure waters of a mountain lake, clear and sparkling in the light of heaven, because kept from a soiling channel, — in this point of view, worthy is the tradition, — how much more worthy to him who looks upon the Bible as the word of God!

In whatever point of view it may be considered, there are strong expressions, there is a force of words, a striking language in the first of Genesis, and if studied, not in verbal criticism, but with a tone of mind in sympathy with its spirit, there may be found light to show the construction of the world and the nature of the elements of which it is composed; and light that will dissipate the mists of skepticism which dim the beauty and glory of creation.

In conclusion, we ask for a few minutes' consideration of the effect of mechanical theories of the universe on faith in the superintending Providence of God. A blameless life and the most fervent piety have often been associated with manifest errors in philosophy; and grievous faults and utter indifference to religion have been accompanied by the most eminent scientific attainments. Therefore a man's theories should never be deemed the test of his religious character. But, on the other hand, the effect of philosophical speculation on the general mind, — its tendency to aid in the formation of the Christian character, or to produce a skeptical spirit, is a proper subject of investigation.

There are no atheists now. No man avows his unbelief; nor are there any theories put forth which do not recognize the existence of the Great First Cause. A purely mechanical theory of the construction of the world, — the nebular hypothesis for instance, acknowledges that God created the material from which the worlds were constructed, and that He gave to it its law, or mode of action. But this being accomplished, the idea of God becomes an abstraction; it is no longer a necessary part of the speculation. Matter has been created, and it acts; the machine has been constructed, and it continues in operation. There is no need of a superintending Providence. Mechanism, the power of matter, stands in its place. The sparrow falls to the ground because of some material sequence. The earth is self-shaped and self-sustained; its relative position and motion are derived from the action of other spheres upon it. A circle is thus described around all material things, excluding God, yet comprehending within its circumference all requisite power for the self-preservation and self-government of matter. Notwithstanding the chilling influence of this theory, a man, from the original bias of his mind, from early education, or by means of the written and spoken word, may retain his faith; but it is not a theory to convert the skeptical, confirm the wavering, or deepen the conviction of any mind. Therefore, although science has enrolled among its votaries the names of many men distinguished for their religious faith, it has not been preëminent as it should have been, for producing and confirming the religious character.

We believe that the ultimate test of the soundness of theory in philosophy should be its religious bearing on the character; and that if any mechanical theory ascribes to matter power to shape for itself its masses, or to determine for itself its position and motion, thereby enabling the mind to regard creation without the idea of a present sustaining God, it is not the true theory. True philosophy will do more than merely to permit or tolerate a belief in God; it will compel the mind to trace all things directly and immediately to Him.

The belief that the universe is a vast machine, in regular constant action because of the inherent powers of matter, arises from a supposed analogy between the works of God and of man. They ought not to be thus associated. The means and processes of the mechanic are not the means and processes of the Artificer of worlds. A steam-engine is constructed; the desired result is to control and change the direction of the force of steam by means of the iron cylinder, of bolts and rods, of pipes and valves. The object is to modify the action of existing force, and this is accomplished by using the known properties of matter, — its elasticity and force of cohesion. The engine constructed, the mechanic stands aside, the spectator of its action. Thus mechanical philosophy regards the world. God, by means of the properties and laws of matter, has constructed a world, and He then becomes only a witness of the action of the perfect machine. But there is no analogy whatever between these processes, or their results. The artisan, by the use of mechanical powers, by wheels, levers, and pullies, seeks to control existing force, — to change established motion. He works with the full understanding, that, in relation to him, there are fixed laws of force, and definite properties in matter. His command over matter comes from the sequences established by a higher power. Not so with God. Matter in relation to Him has no assigned quality, no position derived from its inherent powers. It is clay in the hands of the potter. Its only attribute is obedience, and the form, position, motion given to it, its only possible property. Where it is and what it is are from God. It has no will of its own, nor is the Divine Artificer constrained to apply the law which governs one form of matter to limit the force of another form of matter. Hence to God there is no machinery whatever. A planet is in its place, not by any virtue or energy of its own, nor by the virtue or energy of any other sphere. It is where it is, and moves as it moves, not by a mechanical power, but by a spiritual agency, — not by machinery, but by the will of God. One event does not happen because of another preceding event, but both from His will; one thing is not the cause of another, but all things are obedient.

There is no necessity, arising from inherent properties of matter, that this earth should be the pleasant home of man. Matter of itself cannot minister to human wants. There is no absolute necessity that there should be the regulated succession of phenomena, by which man traces out order and design. There is no necessarily existing connection between material cause and effect. All is from the direct and immediate will of God. The natural sequences of things, the regulated successions of events, are but ministrations to human weakness. The body has wants, and they are provided for; the mind desires knowledge, and the creation is brought down as it were to the level of its comprehension. An order is stamped upon the mind, and to this order, to this low degree of comprehension, is the creation adapted.

A miracle declares the presence of God; but the miracle is only the sudden withdrawal of the veil which uniformity of action spreads over the Supreme Power. And in one point of view even the uniformity of God's Providence is a miracle. For this very regulated succession, and these apparently predetermined results, are the manifestation of the Supreme Power as conformed to the intellectual wants of the individual man. Every instance in which man traces the conformity of one thing to another, every instance of the discovery of order in the heavens or in the earth, is the manifestation of the Divine attributes, which have so constructed the Universe and the soul of man, that the order in the one is the counterpart of the order in the other. In the discovery of this order is the glory of philosophy and its reward, and herein is philosophy taught to bring all its theories to this ultimate test: — Do they impress on the mind, not the idea that there is power in matter, and that the universe is a perfectly constructed machine, but the great idea of a Father in Heaven, who condescends to the weakness of his children, giving them not only day by day their daily bread, but furnishing them, in the evidences of design continually displayed before them, the intellectual strength which may lift their minds to Him?

> "How exquisitely the individual Mind
> (And the progressive powers perhaps no less
> Of the whole species) to the External world
> Is fitted: — and how exquisitely too —
> Theme this but little heard of among men —
> The External world is fitted to the mind;
> And the Creation (by no lower name
> Can it be called,) which they with blended might
> Accomplish — this is our high argument."

We firmly believe that in the progress of time there will remain no mechanical theories of the construction or of the preservation of the universe. Does matter by delegated power, or does God directly determine the place and position of all things? One or the other is the case; there is no divided sway. The universe is a machine with a far-off God; or the present God sustains, upholds, moves every part, and this, not from a law impressed upon matter from the beginning, but because matter has no quality, attribute, or power of its own. It but obeys His present will.

We have attempted to form some definite idea of power, of force, of something which exists independently of matter, which is not a property of matter, and which matter can never wield or direct. We ascribe to the material of creation no power of self-direction or self-control. We believe in power, because we are conscious of exercising it in some degree through the will, by which we can change the position and motion of things. We compare our feeble strength with the force of the elements, and this with the intensity of the motion of the world, and this again with the majestic movements of the extended system of worlds. Nor will the mind pause here; it goes at once to the omnipotence of God. The acknowledgment that the power which, proceeding from human volition, acts directly on matter, is something which is independent of matter, born of the will and not generated by mechanism, is the first step in progress. Comparing this in its utter insignificance with the power which moves the great universe, knowing as we must know that this force can proceed only from mind, we are led upward till we reach the Source of all power.

We look upon the universe as the creation of God, set forth as the symbol of His wisdom, — as His thoughts written out in matter. We are not willing to assign to matter that which is not, and which never can be its property, or to degrade the spiritual into a pantheistic commingling with the material. We cannot conceive of a universe without the idea of the Creator, nor of a universe that can contain God, or be of sufficient extent or value to give scope for the attributes of the Omnipotent. We have no sympathy with theories which present the idea of a gradual induction of order from confusion, of a waiting for completeness, of extended time as requisite to bring about the desired results. We love to think that the worlds came perfect from the hands of their Creator, and that all their changes are but from one perfection to another perfection, without a gradual process, without an intermediate confusion. We believe not in a time when the world was not clothed with perfect wisdom as its vesture. We hold that philosophy to be the most true to the actual construction of things, which most clearly establishes these ideas in the mind. We like to discharge the words, nature, law, regular succession, and the like, and to ascend at once to that which they indistinctly represent, — to the idea of "that pure and holy Mind that with swift thoughts agitates the whole world." Why should we confine ourselves to language which gives the idea of intervening powers, and which thus separates the Deity from His works? The word nature is but a substitution, — the word law is but the obscured representation of Omnipresence, — order, regulated succession, mean Omniscience, — adaptation, harmony, "the agreeable fitness of things," vaguely present the idea of a Father in Heaven.

THE END.

www.ingramcontent.com/pod-product-compliance
Lightning Source LLC
LaVergne TN
LVHW020211110826
845151LV00003B/676

* 9 7 8 1 4 2 5 5 3 4 5 0 9 *